Strategies for Sustainability

The series focuses on "implementation strategies and responses" to sustainability problems – at the organizational, local, national, and global levels.

Our objective is to encourage policy proposals and prescriptive thinking on topics such as: sustainability management, sustainability strategies, lifestyle changes, regional approaches, organisational changes for sustainability, educational approaches, pollution prevention, clean technologies, multilateral treaty-making, sustainability guidelines and standards, sustainability assessment and reporting, the role of scientific analysis in decision-making, implementation of public-private partnerships for resource management, regulatory enforcement, and approaches to meeting inter-generational obligations regarding the management of common resources.

We favour trans-disciplinary perspectives and analyses grounded in careful, comparative studies of practice, demonstrations, or policy reforms. This largely excludes further documentation of problems, and prescriptive pieces that are not grounded in practice, or sustainability studies. Philosophically, we prefer an open-minded pragmatism – "show us what works and why" – rather than a bias toward a theory of the liberal state (i.e. "command-and-control") or a theory of markets. We invite contributions that are innovative, creative, and go beyond the 'business as usual' approaches.

We invite Authors to submit manuscripts that:

- Document and analyse what has and has not worked in practice;
- Develop implementation strategies and examine the effectiveness of specific sustainability strategies;
- Propose what should be tried next to promote greater sustainability in natural resource management, energy production, housing design and development, industrial reorganization, infrastructure planning, land use, business strategy, and organisational changes;
- Prescribe how to do better at incorporating concerns about sustainability into organisations, private action, and public policy;
- Focus on trans-disciplinary analyses grounded in careful, comparative studies of practice or policy reform; and
- Provide an approach "…to meeting the needs of the present without compromising the ability of future generations to meet their own needs," and do this in a way that balances the goal of economic development with due consideration for environmental protection, social progress, and individual rights.

Themes covered in the series are:

Sustainability management
Sustainability strategies
Lifestyle changes
Regional approaches
Organisational changes for sustainability
Educational approaches
Pollution prevention
Clean technologies
Multilateral treaty-making
Sustainability guidelines and standards
Sustainability assessment and reporting
The role of scientific analysis in decision-making
Implementation of public-private partnerships for resource management
Governance and regulatory enforcement
Approaches to meeting inter-generational obligations regarding the management of common resources

Luis Velazquez

Circular Economy and Sustainable Consumption in the Context of SDG 12

Springer

Luis Velazquez
Graduate Program in Sustainability
Department of Industrial Engineering
Interdisciplinary Faculty of Engineering
University of Sonora
Hermosillo, Sonora, Mexico

ISSN 2212-5450 ISSN 2452-1582 (electronic)
Strategies for Sustainability
ISBN 978-3-032-15597-9 ISBN 978-3-032-15598-6 (eBook)
https://doi.org/10.1007/978-3-032-15598-6

This Springer imprint is published by the registered company Springer Nature Switzerland AG
The registered company address is: Gewerbestrasse 11, 6330 Cham, Switzerland

Foreword

There is an abundance of literature on the circular economy (CE); thousands of academic papers and dozens of books had been published on the topic. This proliferation of literature has resulted in conceptual ambiguity and "buzzword fatigue." At this point, one could question whether new CE research is truly advancing knowledge or simply piling onto what Kirchherr has dubbed the "heap of scholarly bullshit."

The CE does matter; it has been hailed as a leading paradigm for sustainability transformation, and is widely endorsed by policymakers, businesses, and scholars alike in addition to now being embedded in the strategies of the EU, China, and dozens of countries, and countless corporate sustainability roadmaps. Conceptually, the CE has achieved remarkable maturity and consensus: frameworks have been refined, principles codified, and roadmaps drawn.

In practice, however, operationalization of circularity remains frustratingly slow and fragmented. After more than a decade of high-level commitments and pilot projects, the gap between theoretical ambition and on-the-ground reality still yawns. The world's economy today is only about 7–8% circular and the global circularity rate has declined in recent years rather than increased, as rising resource use outpaces gains from recycling and reuse.

Why this implementation gap? For years, both research and practice on CE have been overwhelmingly producer centric: circular business models, optimized product designs for longevity, and engineered industrial symbiosis networks. These developments are important but insufficient, because a fully circular economy cannot be achieved by producers alone; the CE ultimately relies on uptake and proper use in the market.

Technology and policy solutions falter when met by indifference, or resistance, from end users. Unless consumer demand evolves in parallel with production, the circular transition will stall. Goods can only circulate as intended if people choose to repair, return, rent, or share them instead of reflexively buying new and discarding. Services like product leasing or take-back only work at scale if consumers accept new notions of access over ownership. Circularity must be desired, chosen, and normalized by consumers in their everyday lives—not just designed for by producers.

Herein lies the justification for new literature on the CE—rather than rehashing and revisiting the same themes, it must focus on underexplored yet critical areas. Consumer-centric issues constitute one of the most significant knowledge gaps in the circular economy literature. We still know far too little about the psychological, social, and cultural mechanisms that might entice individuals to choose a circular option over a linear one—to value a product's story (e.g., recycled or repaired content) over its newness, or to embrace usage models based on sharing and longevity rather than one-off ownership.

What drives someone to pay a premium for a remanufactured good—or conversely, what fears or stigmas hold others back (e.g., concerns about quality, hygiene, or status)? How do social norms and identity play into decisions to buy second-hand fashion, rent tools, or patronize a repair café? What marketing and design cues could make circular offerings more appealing, convenient, or trustworthy to mainstream consumers?

The importance of this missing literature cannot be overstated. Consumer interest and awareness has been pinpointed as the single greatest barrier to circular economy implementation—ahead of technology, finance, or policy issues. Closing that gap is not just an academic exercise—it is arguably the linchpin for translating circular ideals into widespread practice.

The timing for such research also couldn't be more urgent: The initial excitement around sustainability and ESG initiatives has, in many quarters, given way to skepticism and fatigue. Public surveys reveal waning enthusiasm for corporate "green" claims and a collapse in trust: nearly a quarter of consumers in 2025 reported that they "rarely or never" believe companies' sustainability claims.

This book represents a meaningful step toward addressing the gap in consumer-centric literature. It bridges the worlds of production and consumption, linking hard-won insights from circular business practice with fresh understanding of consumer behavior and culture by combining engineering and design thinking with psychology, sociology, and marketing to tackle the human dimensions of circular transition.

The chapters that follow blend conceptual grounding with actionable ideas on how circularity can be made desirable, accessible, and even inevitable in daily life. The authors delve into case studies of successful circular products and services, analyze the behavioral nudges and infrastructural supports that enabled them, and offer strategic guidance to practitioners.

Drawing on evidence from multiple disciplines—from consumer psychology experiments to market trend analyses and digital innovation case studies—the book illustrates how circular consumption can move from niche to norm. Its insights illuminate how companies and policymakers might better align circular offerings with consumer needs and aspirations, whether through new service-based business models, community-based initiatives, or digital platforms that build trust and convenience.

In sum, this work builds a bridge between systemic transformation and individual action—a bridge needed to span the current chasm between circular economy theory and real-world adoption.

The coming 5–10 years will be pivotal. With many nations rolling out "right-to-repair" legislation, product passport schemes, and incentives for take-back and reuse, we have an opening to hardwire circularity into the consumer experience. But policy and industry efforts will succeed only if they resonate with the public's values and day-to-day realities. Key levers—education, transparency, convenience, affordability—must converge to make circular options the attractive default.

This book provides a roadmap for how that convergence can be achieved, offering both vision and practical direction. It is a must-read for anyone serious about accelerating the transition to a genuinely circular economy—one that is not only designed and produced to be circular but also embraced and lived as circular by society at large.

Dr. Julian Kirchherr
Associate Professor
Roskilde University
Roskilde, Denmark

Dr. Mitchell P. Jones
Institute of Material Chemistry and Research
Polymer and Composite Engineering (PaCE) Group,
Faculty of Chemistry
University of Vienna
Vienna, Austria

Preface

This book, *Circular Economy and Sustainable Consumption in the Context of SDG 12*, explores the strategic role of circularity in ensuring sustainable consumption and production patterns. While SDG 12 is often associated with industrial transformation, this book emphasizes that the circular economy and sustainable consumption are not exclusive to industrial systems. Their principles are applicable across all productive sectors—including agriculture, services, construction, tourism, and education—where resource efficiency, waste reduction, and responsible consumption are equally critical. By broadening the scope beyond industry, the book encourages readers to recognize the systemic nature of the circular economy and the need for cross-sectoral collaboration in achieving meaningful progress toward SDG 12 and the broader 2030 Agenda for Sustainable Development.

Written to offer a broad, interdisciplinary framework, the book supports the development of key skills needed to address sustainability challenges across various productive sectors. Rather than focusing on specialized technical abilities, it emphasizes general competencies such as systems thinking, innovation, creativity, digitalization, ethics, and environmental responsibility—core skills for students and professionals aiming to implement circular economy principles in real-world contexts. While faculties of economics, engineering, and sociology play important roles in shaping models of circularity, technical solutions, and social transitions, the complexity of the circular economy and sustainable consumption calls for contributions from natural sciences, humanities, business administration, and computer science. These disciplines are essential for understanding ecological cycles, ethical foundations, organizational change, and digital infrastructures that support circular practices. By promoting systems thinking and lifecycle perspectives, the book seeks to prepare professionals capable of designing strategies to close material and energy loops across sectors, and to support educators, researchers, and policymakers working toward transformative sustainability.

Hermosillo, Mexico — Luis Velazquez

Acknowledgments I am grateful to Zachary Romano, Ph.D., Editor of the book series *Strategies for Sustainability*, and to Rodrigo Lozano, Ph.D., for their kind invitation to contribute to this collection. I also sincerely thank Nora Munguia, D.Sc., and Claudia Jayne Falcón Pérez, M.Sc., for their invaluable support during the development and writing of this book. Their insights, encouragement, and thoughtful input were instrumental in shaping the structure and content of this work. I also appreciate the support from the Industrial Engineering Department, the Interdisciplinary Faculty of Engineering, and the University of Sonora, which provided the institutional backing and the time and space needed to complete this project. My thanks extend to my colleagues and students from the Sustainability Graduate Program, whose perspectives and engagement have enriched this work in meaningful ways. Finally, I thank my family for their unwavering support, patience, and understanding throughout the writing process. Their encouragement has been a source of strength and motivation.

Competing Interests The author has no competing interests to declare that are relevant to the content of this manuscript.

About This Book

This book offers a comprehensive and interdisciplinary exploration of the circular economy paradigm, with a particular emphasis on Sustainable Development Goal 12 (SDG 12): ensuring sustainable consumption and production patterns. Structured in seven interconnected chapters, it guides readers through foundational concepts, strategic frameworks, cultural transformation, enabling technologies, and practical applications across sectors and scales.

Following the Introduction, which situates the book within the broader context of sustainability and educational relevance, Chap. 2 ("The Circular Economy Paradigm and Sustainability Approaches") presents the historical evolution, principles, and systemic nature of circularity, drawing on seminal contributions from Boulding, Stahel, Ayres, and others. Chapter 3 ("Advancing Circular Models in Practice") examines multi-level implementation strategies—macro, meso, micro, and nano—supported by industrial ecology, pollution prevention, and cleaner production frameworks. Chapter 4 ("The Challenge of Downshift Consumerism") addresses the transformation of consumption patterns, emphasizing the role of cultural behavior, generational expectations, and regulatory drivers in shaping sustainable demand. Chapter 5 ("Transforming Organizational Culture for the Circular Economy") explores how leadership, internal education, and stakeholder engagement foster sustainability-oriented cultures, integrating theories such as Open Systems Theory, Contingency Theory, and Institutional Theory. Chapter 6 ("Digitalization and Circular Economy in Industry 4.0") analyzes how digital tools—IoT, AI, blockchain, and digital twins—enable traceability, resource optimization, and lifecycle extension, while critically examining challenges such as e-waste and digital inequality. Chapter 7 ("Applying Digital Solutions for Circular Economy Transformation") presents strategic applications of digital technologies in traceability, waste valorization, energy circularity, modular design, and circular supply chain management, supported by recent empirical studies and policy frameworks.

Designed for undergraduate and graduate courses in sustainability, environmental management, industrial engineering, and interdisciplinary programs, the book emphasizes transversal competencies such as systems thinking, creativity, digital literacy, and ethical responsibility. Chapters 2–5 provide essential theoretical and strategic foundations, while Chaps. 6 and 7 offer applied insights into technological enablers. Professors will find a structured resource to support course design, while policymakers, industry leaders, and educators will benefit from its integration of scholarly research and practical examples.

By bridging theory and practice, this book supports the development of professionals capable of leading circular transitions across sectors, contributing to transformative sustainability beyond the 2030 Agenda.

Contents

1 Introduction 1
References 4

2 The Circular Economy Paradigm 7
2.1 Introduction 8
2.2 Sectoral Drivers of Environmental Degradation 9
2.3 A Multidisciplinary Response to Resource Scarcity and Environmental Collapse 11
2.4 Fundamentals of Circular Economy 13
2.5 Insights on the Social Impact of the Circular Economy 16
2.6 Conclusion 18
References 18

3 Advancing Circular Models in Practice 23
3.1 Introduction 24
3.2 Multi-level Governance and Strategic Alignment for Circularity 24
3.3 Circular Economy from a Systemic Perspective 29
3.4 Circular Economy in Business and Market Dynamics 31
3.5 Tools for Sustainable Consumption in Industrial Ecology, Cleaner Production, and Related Fields. Pollution Prevention 35
3.6 Recycling: A Strategic Enabler for Sustainable Consumption 36
3.7 Conclusion 37
References 38

4 The Challenge of Downshift Consumerism 45
4.1 Introduction 45
4.2 Industrial Consumption and the Global Acceleration of Material Use 46
4.2.1 The Resource Extraction 47
4.2.2 Energy Consumption 49

4.2.3 Waste Generation 51
4.3 Sustainability Drivers of Industrial Consumption Reduction 52
4.3.1 Regulatory and Legal Frameworks 52
4.3.2 Science and Technology 54
4.3.3 Organizational Factors 55
4.3.4 Cultural Factors 57
4.4 The Evolving Role of Industry in Sustainable Consumption 59
4.5 Conclusion 60
References 61

5 Transforming Organizational Culture for the Circular Economy 67
5.1 Introduction 68
5.2 Leadership for Circular Transformation 70
5.3 Generational Leadership for Circular Transformation 73
5.4 Organization-Centered Management Theories 75
5.5 Evolution of Management Theories Toward Circular Economy 76
5.6 Organizational Education as a Circularity Driver 79
5.6.1 Systems Thinking 80
5.6.2 Creativity and Innovation 80
5.6.3 Digital Competence 81
5.6.4 Ethical and Environmental Responsibility 81
5.7 Conclusion 82
References 83

6 Digitalization and Circular Economy in Industry 4.0 89
6.1 Introduction 90
6.2 Enabling Circularity Through Industry 4.0 92
6.3 Applying Digital Solutions for Circular Economy Transformation 94
6.3.1 Traceability and Lifecycle Monitoring 94
6.3.2 Intelligent Material Separation and Waste Valorization 95
6.3.3 Energy Circularity and Carbon Reduction 96
6.3.4 Modular Design and Product Lifecycle Extension 97
6.3.5 Circular Supply Chain Management 97
6.4 Critical Reflections on Digitalization for Circularity 98
6.4.1 Energy Impact of Digital Infrastructure 99
6.4.2 Material Intensity and E-Waste Generation 99
6.4.3 Digital Divides 100
6.4.4 Labor Displacement and Protection in the Era of Industry 4.0 101
6.5 Conclusion 101
References 102

7 Scientific Landscapes of Circular Economy and Sustainable Consumption 107
7.1 Introduction 108
7.2 Methodology 108
7.3 Results 109
7.3.1 General Overview of Research Dynamics 109
7.3.2 Global Research Trends on Circular Economy and Sustainable Consumption 110
7.3.3 Global Research Trends on Circular Economy and SDG 12 112
7.3.4 Global Research Trends on Circular Economy and Materials/Raw Materials 113
7.3.5 Global Research Trends on Circular Economy and Energy 116
7.4 Conclusion 118
References 118

Epilogue 119

About the Author

Luis Velazquez is a Full-Time Professor at the University of Sonora in Mexico. He was the Founding Coordinator of the university's institutional graduate program in sustainability and now serves as the Academic Leader of the Sustainable Engineering Research Group. He holds a Ph.D. in Engineering Sciences with a specialization in Cleaner Production and Pollution Prevention from the University of Massachusetts Lowell, where he also served as an Adjunct Professor in the School of Engineering. With 33 years of experience in academia and research, Dr. Velazquez is recognized as a senior researcher in sustainability, holding the highest level (SNI Level 3) in Mexico's National System of Researchers.

Throughout his career, he has mentored students across multiple countries and contributed extensively to the advancement of sustainability science. His scholarly output includes a substantial number of peer-reviewed articles and several academically regarded books and book chapters published by Springer and other respected international publishers.

Dr. Velazquez has conducted academic stays at renowned institutions, including the Harvard T. H. Chan School of Public Health (USA), Erasmus University Rotterdam (Netherlands), Universidade Paulista (Brazil), and Hochschule Zittau/Görlitz University of Applied Sciences (Germany). He has also served as a guest editor and advisory board member for international journals.

In 2025, he was honored with the Academic Merit Award by the Mexican Association of Faculties and Schools of Engineering (ANFEI), recognizing his lifelong dedication to engineering education and excellence in academic work.

Chapter 1
Introduction

Abstract This book presents the concepts of the circular economy and sustainable consumption as transformative building blocks for achieving Sustainable Development Goal 12 (SDG 12) of the 2030 Agenda. They are much more than technical changes; instead, they represent a radical reengineering of how societies generate, consume, and reuse resources. The chapter traces the intellectual lineage of these concepts, from The Limits to Growth through Industrial Ecology and Cleaner Production, while positioning circularity as a systemic alternative to linear models that externalize waste and resource depletion. It underscores the critical importance of this shift at a time when climate risks are rising, geopolitical tensions are escalating, and global circularity has regressed to 6.9%, while resource use is projected to double by 2060. This book calls upon readers to move beyond incremental efforts and toward integrated, multipronged strategies that promote economic growth while protecting the natural world. It is intended for students, educators, researchers, and policymakers seeking actionable insights and visionary frameworks to lead the circular transition.

Keywords Circular economy · Sustainable consumption · SDG 12 · 2030 Agenda for sustainable development · Resource efficiency · Climate change mitigation · Industrial sustainability

Circular economy and sustainable consumption have gained increasing prominence in global sustainability discourse, particularly within the framework of Transforming Our World: the 2030 Agenda for Sustainable Development. These strategic concepts are essential for achieving Sustainable Development Goal (SDG) 12: ensure sustainable consumption and production patterns, which promotes the reduction of resource depletion, pollution, and waste while supporting climate change mitigation (U.N. General Assembly 2015). Their alignment with SDG 12 underscores their importance in addressing the urgent need to confront harmful industrial practices that contribute to environmental degradation and unsustainable resource use.

The industrial sector, in particular, faces growing scrutiny for its substantial role in climate change and resource depletion. Environmental risks have evolved from abstract concerns to immediate global threats. The Global Risks Report 2024 by

L. Velazquez, *Circular Economy and Sustainable Consumption in the Context of SDG 12*, Strategies for Sustainability, https://doi.org/10.1007/978-3-032-15598-6_1

the World Economic Forum (WEF) ranks extreme weather events as the most likely to trigger catastrophic consequences with widespread impact (WEF 2024). Climate change, a key driver of these events, continues to disrupt global weather systems and intensify their severity (Intergovernmental Panel on Climate Change [IPCC] 2023). Industry alone accounts for 11% of global greenhouse gas emissions, producing 6.5 gigatons of CO_2 annually (United Nations Environment Programme [UNEP] 2024). In the United States, the Environmental Protection Agency (EPA) reported that the industrial sector was responsible for 29.5% of national emissions in 2022 (EPA 2024). These figures reflect the mounting pressure on industries to adopt sustainable practices, driven by consumer expectations, regulatory frameworks, and the imperative to ensure long-term viability in a resource-constrained world (United Nations Industrial Development Organization [UNIDO] 2024).

The foundational ideas behind circular economy and sustainable consumption emerged long before the adoption of Transforming Our World: The 2030 Agenda for Sustainable Development. The seminal report The Limits to Growth (Meadows et al. 1972), published by the Club of Rome, highlighted the finite nature of Earth's resources and the unsustainability of prevailing industrial patterns. Many of the environmental concerns raised in that report remain relevant today, as the global economy continues to grapple with resource depletion, environmental degradation, and unsustainable consumption (Bhardwaj et al. 2025; Kümmel 2024; Mandl 2023).

In response to these challenges, the circular economy emerged as a strategic alternative to linear production models. Its evolution has been shaped by earlier frameworks such as industrial ecology, cleaner production, and pollution prevention, which laid the groundwork for systemic approaches to resource efficiency and waste reduction (Broderick and Usher 2024; Dam et al. 2020; Saavedra et al. 2018). These concepts collectively emphasize the need to rethink material flows, minimize environmental impacts, and transition toward closed-loop systems that align with the goals of SDG 12.

The transition toward sustainable production systems has been shaped by earlier frameworks such as cleaner production and pollution prevention, which promote responsible manufacturing practices and emphasize the protection of natural resources through product and process redesign (Hotta et al. 2021). These approaches have fostered environmental innovation, improved business performance, and laid the foundation for more integrated sustainability strategies, including the adoption of digital technologies in industrial contexts (Fan et al. 2024; Majerník et al. 2023; Xin et al. 2022). Within the 2030 Agenda for Sustainable Development, the guarantee of sustainable consumption and production patterns is central to supporting the well-being of current and future generations. SDG 12 reflects this commitment, and circular economy practices have emerged as key tools for its implementation. Moreover, the integration of circular principles with digital innovation offers new opportunities to advance multiple sustainability targets across sectors (Dantas et al. 2021; Schroeder et al. 2019; Zottele Allende and Nájera Jiménez 2022).

The challenges arising from the COVID-19 pandemic, geopolitical instability, and shifting global priorities have significantly undermined progress toward the Sustainable Development Goals. As nations recalibrate their agendas, the initial momentum of Transforming Our World: The 2030 Agenda for Sustainable Development has weakened (Bhattacharya and Bose 2023). When launched in 2015, the Agenda captured widespread attention across political, business, and academic spheres. However, with only a few years remaining before its deadline, optimism has faded, and key promises—such as "leaving no one behind"—remain largely unmet. Critics argue that the Agenda's goals risk being perceived as aspirational rather than actionable (Weiland et al. 2021). Reports from the High-level Political Forum on Sustainable Development reinforce this concern. According to The Sustainable Development Goals Report 2024, only 17% of assessable targets show sufficient progress (United Nations 2024). Furthermore, current nationally determined contributions (NDCs) fall short of the thresholds set by the Paris Agreement, as many lack comprehensive mitigation strategies to support low-carbon lifestyles (Salem et al. 2021).

Sustainable Development Goal 12, focused on responsible consumption and production, has shown only modest progress: 40% of its targets are advancing, another 40% remain stagnant, and 20% are regressing (United Nations 2024). On the production side, The Circularity Gap Report 2025 highlights a concerning trend—the global circularity rate has declined from 7.2% to 6.9%, indicating that most materials entering the economy are still virgin (Circle Economy 2025). Meanwhile, consumption patterns remain unsustainable: global natural resource use is projected to increase by 60% by 2060 compared to 2020 levels, following a tripling of material use over the past five decades (UNEP 2024).

As the global development agenda continues to evolve, industry-led sustainability initiatives are emerging as pivotal drivers of long-term progress. In response to escalating concerns over resource scarcity, environmental degradation, and climate change, industries have increasingly adopted circular economy and sustainable consumption strategies—such as product take-back programs, material recycling, and remanufacturing—to foster more responsible production practices. These efforts, which predate the 2030 Agenda for Sustainable Development, are expected to persist beyond its horizon. While the circular economy initially emerged from engineering disciplines and was closely associated with industrial processes, growing pressure from consumers, investors, and regulators has encouraged companies to broaden its implementation across the global economic system (Sönnichsen et al. 2024). The service sector, in particular, plays a critical role in this transition due to its intermediary position between producers and consumers (Heyes et al. 2018), prompting the development of circular value propositions and innovative business models centered on sustainability (Pieroni et al. 2019).

Circular economy practices have expanded beyond industrial settings, gaining importance across various sectors. In hospitality, hotels are starting to adopt circular strategies to cut waste and improve resource efficiency (Tiong et al. 2021). Public services are exploring circular methods for water treatment (Guerra-Rodríguez et al. 2020; Tiong et al. 2021), while the construction industry increasingly applies circular principles in building design to lower carbon footprints (Bilal et al. 2020; Tirado et al. 2022). Likewise, the fashion industry is adopting circularity through sustainable design and manufacturing practices (Dissanayake and Weerasinghe 2022), and the food industry is implementing circular strategies to minimize waste and enhance resource use (MacArthur 2019).

Addressing the multifaceted challenges of the circular economy and sustainable consumption demands interdisciplinary collaboration to develop practices that prevent harm, conserve resources, and reduce pollution and waste at the source. Students engaged in these disciplines must cultivate systems thinking, problem-solving abilities, and a deep understanding of circular business models, design principles, supply chain optimization, and environmental impact assessment tools such as lifecycle analysis. Consequently, academic programs across fields—including Economics, Engineering, Business and Management, Environmental Science, Industrial Design, Sociology, Natural Sciences, Humanities, and Computer Science—should embed circular economy and sustainable consumption concepts into their curricula. By doing so, educators can foster the development of professionals equipped to rethink and redesign systems that regenerate resources and minimize energy and material losses.

This book is intended not only for current and future generations of undergraduate and graduate students, but also for educators, researchers, and policymakers committed to advancing transformative sustainability. It aims to support the cultivation of competencies and knowledge essential for leading circular transitions across industries and economic sectors, through integrative approaches that align with the broader goals of systemic change.

References

Bhardwaj E, Alexander R, Becker C (2025) Limits to AI growth: the ecological and social consequences of scaling. In: Computers and society. Association for computing machinery. https://arxiv.org/abs/2501.17980

Bhattacharya R, Bose D (2023) A review of the sustainable development goals to make headways through the COVID-19 pandemic era. Environ Progr Sustain Energ. John Wiley and Sons Inc. https://doi.org/10.1002/ep.14093

Bilal M, Khan KIA, Thaheem MJ, Nasir AR (2020) Current state and barriers to the circular economy in the building sector: towards a mitigation framework. J Clean Product 276. https://doi.org/10.1016/j.jclepro.2020.123250

Broderick N, Usher J (2024) Responsible consumption and production. In: Leal Filho W (ed) Teaching the sustainable development goals to young citizens (10–16 years) (Encycloped.). Springer Nature, Switzerland. https://doi.org/10.4324/9781003232001-18

Circle Economy (2025) The circularity gap report 2025. https://global.circularity-gap.world/

Dam KV, Simeone L, Keskin D, Baldassarre B, Niero M, Morelli N (2020) Circular economy in industrial design research: a review. Sustainability 12(24). https://doi.org/10.3390/su122410279

Dantas TET, de-Souza ED, Destro IR, Hammes G, Rodriguez CMT, Soares SR (2021). How the combination of Circular Economy and Industry 4.0 can contribute towards achieving the sustainable development goals. Sustain Product Consumpt. Elsevier B.V. https://doi.org/10.1016/j.spc.2020.10.005

Dissanayake DGK, Weerasinghe D (2022) Towards circular economy in fashion: review of strategies, barriers and enablers. In: circular economy and sustainability. Springer Nature. https://doi.org/10.1007/s43615-021-00090-5

EPA (2024) Inventory of U.S. Greenhouse Gas Emissions and Sinks 1990–2022. https://www.epa.gov/ghgemissions/inventory-us-greenhouse-gas-emissions-and-sinks-1990-2022

Fan X, Xu Y, Zhang C, Bu Y (2024) How does the location of high-speed railway stations affect the emission reduction effect of industrial structure upgrading? Environ Sci Pollut Res Int 31(4):6106–6124. https://doi.org/10.1007/s11356-023-31615-0

Guerra-Rodríguez S, Oulego P, Rodríguez E, Singh DN, Rodríguez-Chueca J (2020) Towards the implementation of circular economy in the wastewater sector: challenges and opportunities. Water (Switzerland), 12(5). https://doi.org/10.3390/w12051431

Heyes G, Sharmina M, Mendoza JMF, Gallego-Schmid A, Azapagic A (2018) Developing and implementing circular economy business models in service-oriented technology companies. J Clean Prod 177:621–632. https://doi.org/10.1016/j.jclepro.2017.12.168

Hotta Y, Tasaki T, Koide R (2021) Expansion of policy domain of sustainable consumption and production (Scp): challenges and opportunities for policy design. Sustainability (Switzerland) 13(12). https://doi.org/10.3390/su13126763

IPCC (2023) Climate change 2023: synthesis report. contribution of working groups i, ii and iii to the sixth assessment report of the intergovernmental panel on climate change. Atmosphere 13(3):35–115. https://doi.org/10.3390/atmos13030405

Kümmel R (2024) Entropy and the limits to growth. Entropy 26(6). https://doi.org/10.3390/e26060489

MacArthur E (2019) Cities and circular economy for food. In: Ellen macarthur foundation, 1–66. https://www.ellenmacarthurfoundation.org/assets/downloads/Cities-and-Circular-Economy-for-Food_280119.pdf

Majerník M, Chovancová J, Drábik P, Štofková Z (2023) Environmental technological innovations and the sustainability of their development. Ecol Eng Environ Technol 24(4):245–252. https://doi.org/10.12912/27197050/162708

Mandl CE (2023) Managing complexity in social systems. In: Leverage points for policy and strategy (Second.). Cham, Springer. https://doi.org/10.1007/978-3-031-30222-0

Meadows DH, Meadows DL, Randers J, Behrens W (1972) The limits to growth : a report for the Club of Rome's project on the predicament of mankind, 205

Pieroni MPP, McAloone TC, Pigosso DCA (2019) Configuring new business models for circular economy through product-service systems. Sustainability (Switzerland) 11(13). https://doi.org/10.3390/su11133727

Saavedra YMB, Iritani DR, Pavan ALR, Ometto AR (2018) Theoretical contribution of industrial ecology to circular economy. J Clean Prod 170:1514–1522. https://doi.org/10.1016/J.JCLEPRO.2017.09.260

Salem J, Lenzen M, Hotta Y (2021) Are we missing the opportunity of low-carbon lifestyles? international climate policy commitments and demand-side gaps. Sustainability (Switzerland) 13(22). https://doi.org/10.3390/su132212760

Schroeder P, Anggraeni K, Weber U (2019) The relevance of circular economy practices to the sustainable development goals. J Ind Ecol 23(1):77–95. https://doi.org/10.1111/jiec.12732

Sönnichsen S, de Jong A, Clement J, Maull R, Voss C (2024) The circular economy: a transformative service perspective. J Serv Res. https://doi.org/10.1177/10946705241229423

Tiong YY, Sondoh SL, Tanakinjal GH, Iggau OA (2021) Cleaner operations in hotels: recommendation for post-pandemic green recovery. J Clean Prod 283:124621. https://doi.org/10.1016/j.jclepro.2020.124621

Tirado R, Aublet A, Laurenceau S, Habert G (2022) Challenges and opportunities for circular economy promotion in the building sector. Sustainability (Switzerland) 14(3). https://doi.org/10.3390/su14031569

U.N. General Assambly (2015) Transforming our world: the 2030 agenda for sustainable development. A/RES/70/1 15900:1–35

UNEP (2024) Global resources outlook 2024: Bend the trend—pathways to a liveable planet as resource use spikes. International Resource Panel

UNIDO (2024) Informe Sobre El Desarrollo Industrial 2024-Convertir Los Desafíos En Soluciones. https://policycommons.net/artifacts/11754787/informe-sobre-el-desarrollo-industrial-2024/12646014/
United Nations (2024) The sustainable development goals report. https://doi.org/10.4324/9780429282348-52
WEF (2024) Global risks report 2024. https://www.weforum.org/publications/global-risks-report-2024/
Weiland S, Hickmann T, Lederer M, Marquardt J, Schwindenhammer S (2021) The 2030 agenda for sustainable development: transformative change through the sustainable development goals? Politics Govern 9(1):90–95. https://doi.org/10.17645/PAG.V9I1.4191
Xin C, Zheng C, Sun M (2022) Environmental innovation ambidexterity and customer relationship performance: evidence from the manufacturing industry in China. Environ Sci Pollut Res 29(40):60998–61011. https://doi.org/10.1007/s11356-022-20284-0
Zottele Allende AC, Nájera Jiménez LE (2022) Circular economy: contribution to the 2030 agenda. Revista Mexicana de Economia y Finanzas Nueva Epoca 17(4):1–17. https://doi.org/10.21919/remef.v17i4.792

Chapter 2
The Circular Economy Paradigm

Abstract This chapter aims to explore the evolution, principles, and societal implications of the circular economy as a transformative response to resource scarcity and environmental degradation. It begins by contextualizing the emergence of circularity within the broader challenges of global economic growth, climate change, and the limitations of linear production models. Drawing on multidisciplinary perspectives—including economics, engineering, environmental science, and sociology—the chapter examines how traditional industrial systems have contributed to ecological decline through unsustainable resource extraction, waste generation, and energy consumption. The chapter traces the historical development of circular economy concepts, highlighting foundational contributions from scholars such as Kenneth Boulding, Walter Stahel, and Robert Ayres. It emphasizes the systemic nature of circularity, rooted in closed-loop material flows and inspired by natural ecosystems. Key operational principles are discussed, including reuse, recycling, product life extension, and industrial symbiosis, alongside frameworks such as industrial ecology and the Ellen MacArthur Foundation's three pillars of circularity. Furthermore, the chapter addresses the social dimension of circular economy strategies, underscoring the need for equity, community resilience, and inclusive governance. Case studies from various sectors—such as poultry, urban mining, and pulp and paper—illustrate how waste valorization can generate both environmental and social benefits. The chapter concludes by advocating for a holistic approach to circular transformation, one that integrates technical innovation with cultural change, interdisciplinary collaboration, and long-term sustainability goals aligned with SDG 12.

Keywords Circular economy · Circularity · Resource scarcity · Closed-loop systems · Sustainable consumption · Sustainability · SDG 12

L. Velazquez, *Circular Economy and Sustainable Consumption in the Context of SDG 12*, Strategies for Sustainability, https://doi.org/10.1007/978-3-032-15598-6_2

2.1 Introduction

Since the early 1970s, global economic growth has been increasingly shaped by a series of complex and interrelated challenges. Deteriorating economic prospects, the resurgence of protectionist policies, rising geopolitical tensions, and the accelerating effects of climate change (Organisation for Economic Co-operation and Development [OECD] 2025) have all contributed to a noticeable slowdown. Trade barriers and environmental pressures have sparked growing concern about the long-term viability of resource-intensive economic models. Naturally, this is expected to cause most economies to slow down more than they did in 2024, pushing them toward the slowest global growth since 2008 (World Bank 2025). There are also genuine concerns about shortages of raw materials, which are now negatively impacting supply chains (OECD 2022). Furthermore, according to the United Nations Environment Programme (UNEP), global economic growth has relied on extracting and consuming natural resources without regard for sustainable practices, leading to resource exhaustion, environmental degradation, and the phenomenon of climate change (UNEP 2021). More than half of all raw materials ever extracted over the twentieth century were mined between 2000 and 2015, and recent World Bank projections suggest that global demand for virgin raw materials could at least double by mid-century (World Bank 2022). From this perspective, the consensus among economists, policymakers, and sustainability analysts is that current patterns of overconsumption—such as hyperconsumption, planned obsolescence, and intensive production—will further harm macroeconomic growth and cause significant environmental, social, and economic sustainability tensions, which may become irreversible (Qamruzzaman 2024). This issue of resource scarcity is a fundamental challenge in economics. Economics addresses this by examining how to efficiently allocate limited resources through supply, demand, and market mechanisms (Krugman et al. 2022). This concept is closely linked to the factors of production—the inputs society use to produce goods and services that meet societal needs (Landsburg 2012). Traditionally, these factors are divided into three main groups: land, labor, and capital (Xu et al. 2009). According to Semin and Namyatova (2019), land requires special attention because it is finite and non-renewable by nature, encompassing all natural resources, including soil, water, minerals, and ecosystems. Scarcity prevents the free redistribution of rewards for obtaining all desired goods, making scarcity a key concept in economics. The fundamental challenge in economics is balancing unlimited human wants and needs with limited resources (Schettkat 2023). This limitation is one of the greatest threats a production system can face. It underpins the concept of scarcity—a core principle of any economy that forces agents to decide how to allocate resources. Ultimately, all aspects of economics involve considering which goods must be consumed by others, prioritizing choices, and finding ways to utilize resources better to benefit society. More importantly, land resources are scarce not only because of their limited quantity but also because they are highly irreplaceable, especially given their deteriorating condition.

2.2 Sectoral Drivers of Environmental Degradation

The typical business model for most of the twentieth century was an industrial enterprise treated as a "black box"—raw materials went in one end, and finished products for sale came out the other. This model failed to consider the resources used, how waste and pollution are created, and what happens to them afterward. Not only have businesses been blind to the real cost of their environmental impacts by mainly focusing on productivity and profit, but this has also kept the surprisingly wasteful linear economic model going—one that extracts, mass-produces, and discards goods in what is known as business as usual (Egorova 2023). Multiple economic sectors, including intensive agriculture, transportation, energy generation, tourism, and urban expansion, are significant contributors to environmental damage, with industrial production emerging as one of the most substantial sources (Singh et al. 2024). The following paragraphs briefly examine the most significant effects on ecosystems caused by these high-impact sectors, beginning with industry and agriculture.

Industrialization: The pace of change in the Earth's ecological balance has accelerated due to global industrialization, disrupting natural harmony and exacerbating environmental decline worldwide (Hammami et al. 2025). This industrialization usually does not reflect a fundamental prioritization issue—balancing economic goals against environmental preservation (Usman and Balsalobre-Lorente 2022). Major consumers of finite natural resources, including lead, copper, iron, and gold, are in industries that heavily depend on these materials, such as manufacturing, mining, and chemical production. These resources are essential for industrial output but also cause significant contamination of soil, air, and water. Industries extract these resources faster than they can be replenished, and this depletion rate is speeding up. According to UNEP (2024a, b), industrial processes contribute 9% of global greenhouse gas emissions, roughly 6.5 gigatons of CO_2 ($GtCO_2e$), making up 11% of the energy sector's emissions in 2024. Additionally, the industry and energy sector account for 19% of worldwide freshwater withdrawals, with the European Union (49%) and Asia (47%) having the highest levels of industrial water use (United Nations Educational, Scientific, and Cultural Organization [UNESCO] 2023).

Agriculture: The agricultural sector is often scrutinized for its environmental impact, and not without reason. The overuse of chemical fertilizers and synthetic pesticides in intensive farming has caused severe ecological problems, including soil degradation, water pollution, and a decline in biodiversity. Mankhomwa et al. (2022) also point out that the loss of soil quality is only one of the issues caused by fertilizers and pesticides, as they disrupt natural nutrient cycles, damage soil microorganisms, and lead to water pollution through chemical runoff and acidic waste into rivers, lakes, and aquifers. This issue is worsened by land use changes, which often involve converting forests, jungles, or wetlands into farmland—such as through deforestation—causing imbalance at both local and global levels (Albieruti 2024). In fact, by 2023, agriculture was responsible for about 11% of global greenhouse gas emissions (6.5 $GtCO_2e$), and the sector continued to play a significant role in climate change (UNEP 2024a).

Urbanization: Developing urban infrastructure presents a complex set of challenges for urban planners, as it transforms local environments in numerous ways (Al Shawabkeh et al. 2023). Fast-growing cities change land use, increase energy consumption, and raise demand for natural resources. These pressures lead to ecosystem degradation, biodiversity loss, and a significant strain on the environment. The building sector accounted for 6% of global greenhouse gas emissions in 2023 and utilized energy from non-renewable sources, which comprised 36% of the global energy consumption (UNEP 2024b). A significant portion of environmental degradation associated with urban development stems from the materials used in construction. Construction materials, aside from the environmental impacts of their extraction and processing—such as the release of polluting gases—are a primary source of waste. Construction itself consumes a significant amount of energy and generates environmental impacts, as does the production and transportation of cement, steel, or masonry elements (UNEP 2023). This ongoing trend raises important questions about long-term sustainability and highlights the urgent need to shift construction practices toward more responsible models, including innovations in sustainable concrete production and distribution.

Transportation: According to the Intergovernmental Panel on Climate Change—IPCC (2023), the transport sector intersects with nearly all Sustainable Development Goals (SDGs), particularly those related to energy, urban planning, and climate change. However, its broad relevance also presents a significant obstacle to achieving these goals, due to its persistent environmental impacts. Projections indicate that, without substantial intervention, CO_2 emissions from transport could rise by 16 to 50% by 2050, further complicating global sustainability efforts. In 2019, it was responsible for about 8.7 gigatons of CO_2 equivalent—an alarming 75% increase compared to 1990—and accounted for 23% of global energy-related CO_2 emissions. Road transport dominates this footprint, contributing 70% of total emissions, followed by aviation (12%), shipping (11%), and rail (1%). This road transportation remains one of the fastest growing sources of greenhouse gas emissions globally, with emissions rising by more than 2.5% in 2023 (UNEP 2024a). In the European Union, transport activities accounted for 31% of final energy consumption in 2022, with road transport representing the vast majority at 73.6%, followed by water transport (13.0%), air transport (11.4%), and rail transport (1.4%) (European Commission 2024). In the United States, the transportation sector accounted for approximately 30% of total energy consumption in 2022, with gasoline alone comprising 52% of the sector's energy use—highlighting the dominant role of road transportation (U.S. Environmental Protection Agency 2023). In addition to emissions, the automotive sector's value chain also stands out for its large material footprint. During its production process, it consumes significant quantities of natural resources, including steel, aluminum, plastics, glass, titanium, and minerals such as copper and graphite, which are essential for manufacturing electronic components (Sen et al. 2019).

Energy Production: Over the past few years, energy production has been a significant concern, especially with fossil fuels still playing a disproportionately large role in a global economy that needs to be decarbonized to stay within environmental limits. Still, coal, oil, and natural gas are used to generate much of the world's electricity—not only causing toxic air pollution but also being a significant driver of climate change. In 2024, fossil fuels will still account for nearly 59% of global electricity generation, with coal contributing more than a third (34.5%) and remaining ahead of oil, although its share is declining (24.6%). In third place, natural gas contributes approximately 22% of generation (Low Carbon Power 2024). Emissions from the electricity sector—estimated at 15.1 gigatons of CO_2 equivalent *(Gt* CO_2e*)*—are the largest source of global greenhouse gas (GHG) emissions in 2023 (UNEP 2024a).

The data shows the ongoing reliance on fossil fuels and limited natural resources, along with the environmental effects of various economic sectors studied. It emphasizes the need to speed up the shift to circular economy practices.

2.3 A Multidisciplinary Response to Resource Scarcity and Environmental Collapse

Concerns about the sustainability of current economic systems—mainly linear resource extraction, large amounts of waste and leftovers, and the ongoing depletion of finite raw materials—have been expressed by industry leaders for decades. In 1972, a group of scientists at the Massachusetts Institute of Technology highlighted this issue in the Limits to Growth Report, which argued that exponential increases in non-renewable resource use, pollution, population, industrial output, and food production would ultimately lead to a collapse of the global economy (Turner 2008). This was the first international scientific report to warn that human beings were at risk of overusing finite resources and that if those trends persisted, the global productive system could break under pressure. Therefore, the report, which sparked significant controversy due to its alarming implications, also emphasized something science already knew: non-renewable resources are not only limited but could ultimately lead to the complete collapse of economic systems (Jaeger 2019).

Since then, governments have faced the challenge of turning scientists' groundbreaking analyses and warnings about the consequences of unchecked economic growth and population increase into practical solutions (Al-Refai 2020). Current concerns about the physical limits to economic growth encompass a vision that is not just rooted in basic economics, but also includes truly transdisciplinary methods (Catlow et al. 2020). These transdisciplinary methods have diverse applications across fields as wide ranging as engineering, environmental science, sociology, public policy, psychology, and education (Dimitrova et al. 2022). Applied science and engineering utilize natural knowledge to develop products, services, and systems that can repeatedly utilize limited materials (Brecher and Weck 2022), thereby providing value for sustainable industrialization over time (Rosen 2012). Human environmental science examines the effects of human activities on the environment and natural resources and works to develop strategies for conserving biodiversity and ecosystems

(Brecher and Weck 2022). Public policy addresses finite resources through ground rules that ensure their sustainable maintenance (Mahardhani 2023). Preparing future generations for a greener and more resilient society underscores the essential role of environmental education (Boca and Saraçli 2019; Hadjichambis and Paraskeva-Hadjichambi 2020). Moreover, students' participation in conservation efforts further supports cultivating environmental awareness and active participation in the better use of natural resources (Natori et al. 2025). On the other hand, it is well known that cultural norms and values significantly influence how resources are consumed, accessed, and allocated (Longo et al. 2021). Likewise, psychology has contributed studies on the determinants of resource use; although fewer in number, these works increasingly focus on actions to protect the socio-environmental system (Corral Verdugo 2012). Together, these disciplines can also help explain twenty-first-century consumerism, which drives the rate of purchasing goods and services above what is needed to meet human subsistence, putting additional pressure on increasingly scarce "land resources" and many non-renewable natural resources (Bellemare et al. 2022).

Unfortunately, due to the lack of an integrative framework, each discipline considers the scarcity of natural resources based on its own theoretical frameworks and priorities, sometimes emphasizing different aspects of environmental constraints on human economic growth (Abbot and Malani 2025). In response to the complex and interconnected nature of these challenges, investigations into sustainability science have been launched, not so much as a "new" field but as an integrated discipline that combines knowledge bases and practices, from global to local perspectives, across natural and social sciences, engineering, and medical sciences (Clark and Dickson 2003). Such developments have prompted a deeper reflection on how societies produce, consume, and manage waste, laying the groundwork for alternative approaches, such as the circular economy, which seeks to reconcile economic activity with ecological limits.

In a world of rapid change, the circular economy has emerged as a solution—a new way to conserve natural resources through reuse and recycling while also boosting economic activity. Sustainable consumption plays a vital role in this approach, encouraging a shift toward production systems that rely less on finite resources and produce less waste. Rather than being just a theoretical concept, this approach reflects a growing global awareness of the consequences of unsustainable supply chains and consumption models. The Global Resources Outlook 2024 highlights that the relentless use of resources as a core driver is causing the "triple planetary crisis": climate change, loss of nature and biodiversity, and pollution and waste. It emphasizes the overexploitation of natural resources as a primary factor and urges different economic stakeholders to adopt circular economy principles as a key strategy for managing natural resources sustainably. This structural shift is seen as essential to halt, or at least slow down, current levels of resource extraction, which are projected to increase by 60% over 2020 levels by the year 2060 (UNEP 2024b).

2.4 Fundamentals of Circular Economy

The circular economy is an economic model that strategically and sustainably tackles the challenge of improving resource efficiency and waste management. It encourages reuse extensively by reducing reliance on primary raw materials; minimizing waste; extending the lifespan of products and materials; and establishing closed-loop systems for goods, components, or collateral materials (Ghisellini et al. 2016). Due to the slow, fragmented, and non-linear development of the circular economy, setting a clear timeline for its progress is challenging. One reason for this complexity is that issues like the overuse of natural resources and recycling were recognized long before the circular economy became a focus among environmental scientists, economists, societal groups, and policymakers. However, these issues were poorly understood at the time. Historically, traditional societies have practiced circularity in various ways, such as crop rotation in pre-Columbian civilizations and reusing materials like papyrus and flax in ancient Egypt. The latter offers early examples of conserving natural resources (Tuladhar et al. 2021). Aristotle also noted that natural resources needed to be preserved in a world governed by scarcity, where production and consumption must aim for stability (Rivera González 2024).

While the modern concept of the circular economy doesn't have a specific start date, its significance began in the late 1970s, well before it was officially recognized by scholars, policymakers, and businesses (Espinoza 2023). The circular economy, as it is understood today, is the result of a collective effort by researchers, policymakers, and practitioners, who have added new layers of meaning and application over many years. In the academic community, it is often connected to earlier ideas from various scholars. Kenneth Boulding was among the first to introduce the idea of circularity in his famous essay "The Economics of the Coming Spaceship Earth" (Boulding 1966). The term "spaceship economy" was used by him to illustrate that natural resources would eventually run out. Simultaneously, another key figure, Walter R. Stahel, advanced the concept with his research on closed-loop industrial cycles, known as "cradle to cradle." Along with Genevieve Reday, Stahel authored a 1976 report for the European Commission titled "The potential of substituting manpower for energy," which emphasized extending product life through maintenance, reuse, repair, and remanufacturing (Reday-Mulvey 1976). In the early 1980s, he expanded on these ideas in his paper, "The Product-life Factory," in which he expresses valid arguments for shifting from linear to circular manufacturing models (Stahel 1982).

The concept of circularity also stems from how natural ecosystems function, utilizing energy and materials that circulate in closed loops to produce minimal waste. This analogy to natural-system metabolism evokes the concept of industrial metabolism, introduced in the late 1980s by Robert Ayres, who examined how material and energy flow through industrial systems, much like biological systems. He later published his work in numerous journals, including the textbook Industrial Metabolism: Restructuring for Sustainable Development (Ayres and Simonis 1994), which was one of the first introductory texts on industrial ecology. Building on this biomimetic idea, Robert Frosch and Nicholas E. Gallopoulos expanded the concept, proposing that manufacturing should imitate nature's cyclic flows (Frosch and Gallopoulos 1989). Their notion of an "industrial ecosystem" marked a breakthrough in viewing industry as a network of interconnected operations, similar to natural systems. Around the same time, economists David Pearce and R. Kerry Turner defined the circular economy in their 1990 book, "Economics of Natural Resources and the Environment" (Pearce and Turner 1990). They explained the economic reasons for creating material loops—such as reuse, recycling, and waste reduction—alongside the ecological ideas presented by Frosch and Gallopoulos. These works laid much of the theoretical groundwork that later underpinned the development of industrial ecology as an academic discipline, exemplified by Thomas Graedel and Braden Allenby's book "Industrial Ecology" (Graedel and Allenby 1995). This vision echoed Walter Stahel's early concept of "from cradle to cradle," a strategy designed for a circular economy based on principles of durability, reusability, and system closure. The idea gained wider popularity in the 2000s with the publication of William McDonough and Michael Braungart's book, "Cradle to Cradle: Remaking the Way We Make Things," which advocated for product designs that imitate nature's circular processes and generate no waste (McDonough and Braungart 2002).

These foundational contributions have collectively fostered a model that has significantly influenced the adoption of circular economy principles, initially in industrial sectors and particularly in manufacturing (Bhattacharya 2021). Over the following decades, it has evolved into a practical model applicable across various fields, including agriculture, urban planning, education, fashion, and public policy. By the late twentieth and early twenty-first centuries, many scholars and organizations adopted diverse perspectives on circularity, turning it into a discourse that shapes mainstream development (Ugalde Hernández 2021).

Decoupling economic growth from unsustainable resource use is an increasing concern that is more and more reflected in government agendas. While some level of relative decoupling is happening, it remains insufficient to achieve the sustainable development goals. This is highlighted by UNEP (2011), which shows the scale of the problem with figures indicating that global resource extraction, both renewable and non-renewable, rose from 40 billion tons in 1980 to 60 billion tons in 2005, and could reach an astonishing 100 billion tons by 2050 if current trends continue. Follow-up analyses support this trend, with a modeling review suggesting that by 2050, between 320 and 849 million hectares of natural land could be converted to cropland under business-as-usual scenarios, resulting in significant biodiversity loss and negative impacts on ecosystem services (UNEP 2014). Global estimates also indicate that resource extraction and processing are responsible for approximately 90% of biodiversity loss and water stress, as well as nearly 50% of global greenhouse gas emissions (UNEP 2019).

It is essential to recognize that the concept of a circular economy encompasses not only traditional waste management but also a broader systemic transformation. As reported by Kirchherr et al. (2017), the body of scientific literature reports a wide range of interpretations in 114 published definitions across 17 dimensions. Such diversity fosters ongoing discussions that prevent uniformity and complacency, thereby maintaining a dynamic dialogue that allows for context-specific innovation instead of rigid agreement. Although the well-known adage "reduce, reuse, recycle" remains a core principle, the expanding range of interpretations calls for an interrogation of the transformative potential attributed to the circular economy (Shah 2016). Nevertheless, the desire for a single, clear, and prescriptive definition still exists among sectors seeking clarity and practical guidance (Schulz et al. 2019).

Importantly, circularity is not applied uniformly; it becomes operational at different levels depending on the specific roles, responsibilities, and capacities of the actors involved, as argued by Korhonen et al. (2017) operationalizing the principles of the circular economy, it is not a straightforward task due to the many challenges that supporters must face, such as thermodynamic limitations on accessing recoverable energy and materials, establishing system boundaries for complex planetary supply chains, and rebound effects, where gains in efficiency can lead to increased consumption. The Ellen MacArthur Foundation, a global leader in promoting the circular economy, describes this model through three main principles: eliminate waste and pollution, keep products and materials in use at their highest value, and restore nature (Ellen MacArthur Foundation 2024). Other scholars, such as Suárez-Eiroa et al. (2021), suggest 12 operational principles for the full implementation of the circular economy concept. Together, these principles reflect a multidimensional approach to implementing the circular economy, underscoring the need for systemic, integrated strategies that span the domains of sustainability, circularity, and resilience.

In the sustainability domain:

P1: Adjust the inputs to the socioeconomic system.
P2: Adjust the outputs from the socioeconomic system.
P3: Reduce the general needs for resources.
In the circularity domain:
P4: Close the system.
P5: Maintain the value of the resources within the system as much as possible.

In the resilience domain:

P6: Educating for a sustainable and circular economy.
P7: Design for a sustainable and circular economy.
P8: Maintain diversity and redundancy.
P9: Manage connectivity.
P10: Manage slow variables and feedback.
P11 Foster collaborative polycentric governance.
P12 Broader active participation and transparency.

Reframing the domain-based organization of these 12 principles through the lens of macro-, meso-, and micro-levels of intervention offers a more actionable perspective for implementation:

Macro-level interventions:

P1: Adjust the inputs to the socioeconomic system.
P2: Adjust the outputs from the socioeconomic system.

P3: Reduce the general needs for resources.
P6: Educating for a sustainable and circular economy.
P10: Manage slow variables and feedback.
P11 Foster collaborative polycentric governance.
P12 Broader active participation and transparency.

Meso-level interventions:

P4: Close the system.
P5: Maintain the value of the resources within the system as much as possible.
P7: Design for a sustainable and circular economy.
P9: Manage connectivity.
P8: Maintain diversity and redundancy.
Micro-level interventions:
P3: Reduce the general needs for resources.
P5: Maintain the value of the resources within the system as much as possible.
P6: Educating for a sustainable and circular economy.

Applying later principles of circularity relies on actions at different spatial levels, from individual choices to global governance. Most principles can be applied at any scale, especially at the micro-level, where companies have more flexibility to incorporate them into their sustainability policies.

2.5 Insights on the Social Impact of the Circular Economy

In this chapter, the emergence of the circular economy is elaborated upon as more than just an environmental and resource-related strategic response; it is also seen as a logical next step in industrial production methods, corporate social responsibility, and corporate sustainability. Initial measures focused on reorganizing supply chains and improving operational efficiency—both of which are fundamental elements in the development of closed-loop systems designed to minimize waste and maximize resource reuse (Genovese et al. 2017). Actually, waste streams have become valuable resources to an increasing number of industrial firms, and some have begun reengineering their handling (Okon Recycling 2025). One consequence of this transition was the rising trend of materials once considered waste—such as organic residues, industrial byproducts, and urban waste—being efficiently valorized and reused as raw materials (Chavan et al. 2022). In fact, these companies demonstrated how environmental responsibility could indeed foster innovation and profitability through the innovative reuse of waste as a legitimate input, setting an example for the resilience and resource efficiency crucial to circular economy industrial models. For example, the poultry industry utilizes waste products, including eggshells, waste feathers, and broiler byproducts, to produce bioenergy, organic fertilizers, and nutrient-rich feed components through methods such as anaerobic digestion, pyrolysis, and gasification, thereby converting waste into wealth (Kanani et al. 2020). Urban mining—the recovery of raw materials from municipal solid waste—has also emerged as a viable solution. Waste streams, such as E-waste, plastics, and construction and demolition waste, are now processed to become part of the manufacturing cycle, fostering more circular industrial systems (Tejaswini et al. 2022). The pulp and paper sector provides additional examples of waste valorization, with bio-sludge, fly ash, and lime mud being utilized to produce green energy, sorbents, or even construction clinkers (Kumar and Verma 2024).

Innovation in waste management is increasingly driven by societal demands for environmental justice, public health, and inclusive economic growth (Das et al. 2019). Case studies from the poultry, urban mining, and pulp and paper industries illustrate how waste valorization practices generate not only environmental benefits but also significant social advantages. In urban contexts, recovering materials from solid waste reduces exposure to harmful pollutants and creates employment through informal recycling initiatives (Kumar et al. 2023). In the pulp and paper sector, improved waste management enhances safety standards and fosters economic opportunities for surrounding communities. Similarly, the poultry industry's use of organic waste for bioenergy and fertilizers contributes to cleaner operations and regional resilience (Kanani et al. 2020; Kumar and Verma 2024; Tejaswini et al. 2022).

Despite these promising developments, scholars have increasingly highlighted the lack of progress in addressing the social dimension of circular economy strategies. Paunovic et al. (2023) emphasize the need to prioritize equity, participation, and community resilience. Corvellec et al. (2022) argue that dominant ideological and economic narratives depoliticize sustainability and obscure its uncertain contribution to broader goals. Geissdoerfer et al. (2017) note that the relationship between consumer behavior, labor conditions, and corporate actions remains unclear, particularly regarding equity and community well-being. Murray et al. (2017) call for reinvestment in social value, while Zavos et al. (2024) critique the dominance of technical and managerial narratives that neglect cultural, political, and ethical considerations. Scheel and Aguiñaga (2025) advocate for systemic transitions rooted in place-based engagement and long-term resilience. Mies and Gold (2021), through causal loop analysis, reveal that working conditions, human rights, and community well-being are often overlooked. Skene (2022) further warns that circular economy models focused on resource cycling and industrial optimization lacks the systemic scope needed to address complex socio-environmental challenges, proposing instead an embedded economy grounded in Earth system dynamics and ecological resilience. In addition, a criticism of technology claims that technological developments are often concentrated in urban or industrial zones, leaving rural areas largely untouched (Barkin 1989). Consequently, applying circular technologies and innovations may only create territorial divides in some regions, especially when rural or peripheral areas are overlooked, making it harder for certain groups to benefit from such development. In many countries, a significant number of recycling and reuse jobs are performed by the informal sector (U.S. Environmental Protection Agency 2025), highlighting the importance of integrating these jobs into circular chains that offer good working conditions and social recognition.

2.6 Conclusion

The circular economy has evolved from a conceptual response to resource scarcity into a multidimensional framework for sustainability. This chapter has examined its historical development, sectoral impacts, and operational principles, highlighting the urgency of transitioning from linear models to regenerative systems. While circularity has gained visibility in industrial and policy spheres, its implementation often remains focused on technical efficiency, leaving aside the social dimensions that are essential for meaningful transformation.

To fulfill its potential, circular economy strategies must be designed with a clear commitment to equity, participation, and community resilience. Waste valorization practices, when thoughtfully implemented, can contribute not only to environmental restoration but also to public health, employment, and regional development. However, these benefits will remain limited unless circularity is embedded in inclusive governance models and supported by cultural shifts that recognize the human dimension of sustainability.

Understanding the circular economy as a societal transition—rather than a purely technical solution—requires a grounded, interdisciplinary approach. Labor conditions, informal economies, and territorial disparities must be addressed as integral components of circular strategies. Only by aligning circularity with long-term well-being, justice, and ecological integrity can we ensure that its implementation contributes to a truly sustainable future.

References

Abbot DS, Malani A (2025) Revisiting the physical limits to economic growth, with a focus on the waste heat limit. PLoS ONE 20:1–21. https://doi.org/10.1371/journal.pone.0319217

Al-Refai M (2020) The impact of e-government on economic growth in GCC countries. Int Rev (1–2). https://doi.org/10.5937/intrev2001018a

Al shawabkeh R, AlHaddad M, al_fugara A, Arar M, Alhammad R, Alshraah M, Alhamouri M (2023) Toward sustainable urban growth: Spatial modeling for the impact of cultural and natural heritage on city growth and their role in developing sustainable tourism. Alexandria Eng J 69. https://doi.org/10.1016/j.aej.2023.02.001

Albieruti S (2024) The impact of agriculture on the consumption of natural resources (An analytical study). Int J Res Stud Publish 5(59):163–177. https://doi.org/10.52133/ijrsp.v5.59.7

Ayres RU, Simonis UE (1994) Industrial metabolism : restructuring for sustainable development. In: Robert UA, Udo ES (eds) The greening of industrial ecosystems, pp 21–37. https://digitallibrary.un.org/record/162827

Barkin D et al (1989) Adaptación de tecnologías al medio rural en un entorno sustentable 53:160. https://andoni.garritz.com/documentos/Lecturas.CS. Garritz/Ciencia.y.Sustentabilidad/Tecnologias.Medio.Rural.pdf

Bellemare MF, Martin-Déry S, Ziegler R, Vézina M, Raufflet E, Walsh A (2022) Synergizing social economy and circular economy. Canad J Nonprofit Soc Econ Res 13(1). https://doi.org/10.29173/cjnser591

Bhattacharya A (2021) Achieving sustainability in supply chain operations in the interplay between circular economy and Industry 4.0. Product Plann Contr 34(10):867–869. https://doi.org/10.1080/09537287.2021.1981032

Boca GD, Saraçli S (2019) Environmental education and student's perception, for sustainability. Sustainability (Switzerland) 11(6). https://doi.org/10.3390/su11061553

Boulding K (1966) The economics of the coming spaceship earth. Environmental quality in a growing economy, resources for the future. Johns Hopkins University Press, Baltimore, pp 3–34

Brecher C, Weck M (2022) Engineering. In: Lecture notes in production engineering, vol Part F1134. Springer Nature, pp 639–677. https://doi.org/10.1007/978-3-658-34622-5_15

Catlow CR, Davidson M, Hutchings GJ, Mulholland A (2020) Science to enable the circular economy: Science to enable the circular economy. Philosophic Trans Royal Soc A: Math Physic Eng Sciences. Royal Society Publishing. https://doi.org/10.1098/rsta.2020.0060

Chavan S, Yadav B, Atmakuri A, Tyagi RD, Wong JWC, Drogui P (2022) Bioconversion of organic wastes into value-added products: a review. Biores Technol 344:126398. https://doi.org/10.1016/J.BIORTECH.2021.126398

Clark WC, Dickson NM (2003) Sustainability science: the emerging research program. In: Proceedings of the national academy of sciences of the United States of America. https://doi.org/10.1073/pnas.1231333100

Corral Verdugo V (2012) The positive psychology of sustainability. Environ Dev Sustain 14(5):651–666. https://doi.org/10.1007/s10668-012-9346-8

Corvellec H, Stowell AF, Johansson N (2022) Critiques of the circular economy. J Ind Ecol 26(2):421–432. https://doi.org/10.1111/JIEC.13187

Das S, Lee SH, Kumar P, Kim KH, Lee SS, Bhattacharya SS (2019) Solid waste management: scope and the challenge of sustainability. J Clean Prod 228:658–678. https://doi.org/10.1016/J.JCLEPRO.2019.04.323

Dimitrova T, Ilieva I, Angelova M (2022) Exploring factors affecting sustainable consumption behaviour. Administrat Sci 12(4). https://doi.org/10.3390/admsci12040155

Egorova E (2023) Linear to circular: an alternative model of economic development. Positiv Changes 49–55. https://doi.org/10.55140/2782-5817-2023-3-s1-49-55

Ellen MacArthur Foundation (2024) Circular economy principles. https://www.ellenmacarthurfoundation.org/circular-economy-principles. Accessed 17 July 2025

Espinoza HA (2023) Economía circular: una aproximación a su origen, evolución e importancia como modelo de desarrollo sostenible. Revista de Economía Institucional 25(49):109–134. https://doi.org/10.18601/01245996.v25n49.06

European Commission (2024) Final energy consumption in transport-detailed statistics. https://ec.europa.eu/eurostat/statistics-explained/index.php?title=Final_energy_consumption_in_transport_-_detailed_statistics (September 2024), 1–16.

Frosch RA, Gallopoulos NE (1989) Strategies for manufacturing. Sci Am 261(2):144–153

Geissdoerfer M, Savaget P, Bocken NMP, Hultink EJ (2017) The circular economy—a new sustainability paradigm? J Clean Prod 143:757–768. https://doi.org/10.1016/j.jclepro.2016.12.048

Genovese A, Acquaye AA, Figueroa A, Koh SCL (2017) Sustainable supply chain management and the transition towards a circular economy: evidence and some applications. Omega 66:344–357. https://doi.org/10.1016/J.OMEGA.2015.05.015

Ghisellini P, Cialani C, Ulgiati S (2016) A review on circular economy: the expected transition to a balanced interplay of environmental and economic systems. J Clean Prod 114:11–32. https://doi.org/10.1016/j.jclepro.2015.09.007

Graedel TE, Allenby BR (1995) Industrial ecology. Prentice Hall College Div. https://archive.org/details/industrialecolog0000grae_j8q9. Accessed 17 July 2025

Hadjichambis AC, Paraskeva-Hadjichambi D (2020) Education for environmental citizenship: the pedagogical approach 237–261. https://doi.org/10.1007/978-3-030-20249-1_15

Hammami S, Bachegour H, Samour A, Dimnwobi SK, Mati S (2025) Technological innovation, industrialization, natural resources and ecological quality: evidence from the load capacity factor. Discov Sustain 6(1):1–17. https://doi.org/10.1007/S43621-025-01074-3/METRICS

IPCC (2023) Transport. Clim Change 2022-Mitigat Clim Change 1049–1160. https://doi.org/10.1017/9781009157926.012

Jaeger C (2019) Limits to growth. Encycloped Ecol 1–4(4), 367–369. https://doi.org/10.1016/B978-0-444-63768-0.00630-2

Kanani F, Heidari MD, Gilroyed BH, Pelletier N (2020) Waste valorization technology options for the egg and broiler industries: a review and recommendations. J Clean Prod 262:121129. https://doi.org/10.1016/J.JCLEPRO.2020.121129

Kirchherr J, Reike D, Hekkert M (2017) Conceptualizing the circular economy: an analysis of 114 definitions. Resour Conserv Recycl Elsevier B.V. https://doi.org/10.1016/j.resconrec.2017.09.005

Korhonen J, D'Amato D, Droste N, Allen B, Kettunen M, Lähtinen K et al (2017) Green, circular, bio economy: a comparative analysis of sustainability avenues. J Clean Prod 168:716–734. https://doi.org/10.1016/J.JCLEPRO.2017.09.053

Krugman PR, Wells R, Graddy K (2022) Fundamentals of Economics. (Reverté, Ed.) (4ta ed.). https://www.reverte.com/libro/fundamentos-de-economia-4a-ed-5a-ed-original_140754/. Accessed 10 June 2025

Kumar A, Singh E, Mishra R, Lo SL, Kumar S (2023) Global trends in municipal solid waste treatment technologies through the lens of sustainable energy development opportunity. Energy 275:127471. https://doi.org/10.1016/J.ENERGY.2023.127471

Kumar V, Verma P (2024) Pulp-paper industry sludge waste biorefinery for sustainable energy and value-added products development: a systematic valorization towards waste management. J Environ Manage 352:120052. https://doi.org/10.1016/J.JENVMAN.2024.120052

Landsburg SE (2012) The armchair economist. New York. https://books.google.com.mx/books/about/The_Armchair_Economist.html?id=BWe2AAAAIAAJ&redir_esc=y

Longo SB, Isgren E, Clark B, Jorgenson AK, Jerneck A, Olsson L et al (2021) Sociology for sustainability science. Discov Sustain 2(1). https://doi.org/10.1007/s43621-021-00056-5

LowCarbonPower (2024) Low-carbon power: monitor the growth of low carbon energy. https://lowcarbonpower.org/

Mahardhani AJ (2023) The role of public policy in fostering technological innovation and sustainability. J Contemp Administr Manag (ADMAN) 1(2):47–53. https://doi.org/10.61100/adman.v1i2.22

Mankhomwa J, Tolhurst R, M'biya E, Chikowe I, Banda P, Mussa J et al (2022) A qualitative study of antibiotic use practices in intensive small-scale farming in urban and peri-urban Blantyre, Malawi: implications for antimicrobial resistance. Front Vet Sci 9. https://doi.org/10.3389/fvets.2022.876513

McDonough W, Braungart M (2002) Remaking the way we make things: cradle to cradle. North Point Press, New York 45. https://books.google.com/books/about/Cradle_to_Cradle.html?id=KFX5RprPGQ0C. Accessed 17 July 2025

Mies A, Gold S (2021) Mapping the social dimension of the circular economy. J Clean Prod 321:128960. https://doi.org/10.1016/J.JCLEPRO.2021.128960

Murray A, Skene K, Haynes K (2017) The circular economy: an interdisciplinary exploration of the concept and application in a global context. J Bus Ethics 140(3):369–380. https://doi.org/10.1007/s10551-015-2693-2

Natori Y, Nakagawa T, Saito T, Kato T, Seo T, Yata M et al (2025) Conservation education for sustainable development through field internship at NGOs. Curr Res Environ Sustain 9:100278. https://doi.org/10.1016/J.CRSUST.2024.100278

OECD (2022) The supply of critical raw materials endangered by Russia'S War on Ukraine. https://www.oecd.org/en/publications/the-supply-of-critical-raw-materials-endangered-by-russia-s-war-on-ukraine_e01ac7be-en.html

OECD (2025) Economic outlook, interim report steering through uncertainty. Paris. https://doi.org/10.1787/89af4857-en

Okon Recycling (2025) Waste stream management: a complete guide for businesses and communities. Consumer Recycling Initiatives. https://www.okonrecycling.com/consumer-recycling-initiatives/learn-about-recycling/waste-stream-management/. Accessed 21 August 2025

Paunovic I, Müller C, Deimel K (2023) Citizen participation for sustainability and resilience: a generational cohort perspective on community brand identity perceptions and development priorities in a rural community. Sustainability (Switzerland), 15(9). https://doi.org/10.3390/su15097307

Pearce DW, Turner RK (1990) Economics of natural resources and the environment. The Johns Hopkins University Press. https://doi.org/10.2307/1242904

Qamruzzaman M (2024) Do natural resources bestow or curse the environmental sustainability in Cambodia? Nexus between clean energy, urbanization, and financial deepening, natural resources, and environmental sustainability. Energ Strat Rev 53:101412. https://doi.org/10.1016/J.ESR.2024.101412

Reday-Mulvey G (1976) The potential for substituting manpower for energy. Commission of the European Communities. https://openlibrary.org/books/OL3867197M/The_potential_for_substituting_manpower_for_energy. Accessed 21 August 2025

Rivera González M (2024) Economía Según Aristóteles: Visión y Principios. Economía Histórica. https://economiahistorica.com/economia-segun-aristoteles/. Accessed 22 May 2024

Rosen MA (2012) Engineering sustainability: a technical approach to sustainability. Sustainability 4(9):2270–2292. https://doi.org/10.3390/su4092270

Scheel C, Aguiñaga E (2025) A systems view of circular economy. Sustainability (Switzerland) 17(3). https://doi.org/10.3390/su17031268

Schettkat R (2023) Fundamentals of economics. In the Behavioral Economics of John Maynard Keynes. https://doi.org/10.4337/9781802204896.00007

Schulz C, Hjaltadóttir RE, Hild P (2019) Practising circles: studying institutional change and circular economy practices. J Clean Product 237. https://doi.org/10.1016/j.jclepro.2019.117749

Semin AN, Namyatova LE (2019) Land as a factor of production in agriculture and features of agricultural practices. Int J Mech Eng Technol 10(2)

Sen B, Onat NC, Kucukvar M, Tatari O (2019) Material footprint of electric vehicles: a multiregional life cycle assessment. J Clean Prod 209:1033–1043. https://doi.org/10.1016/J.JCLEPRO.2018.10.309

Shah V (2016) The circular economy's hidden wealth. Eco-Business. https://www.eco-business.com/news/the-circular-economys-hidden-wealth/. Accessed 3 September 2025

Singh S, Deep Sharma G, Radulescu M, Balsalobre-Lorente D, Bansal P (2024) Do natural resources impact economic growth: an investigation of P5 + 1 countries under sustainable management. Geosci Front 15(3). https://doi.org/10.1016/j.gsf.2023.101595

Skene KR (2022) The circular economy: a critique of the concept. CSR Sustain Ethics Govern 99–116. https://doi.org/10.1007/978-3-030-94293-9_5

Stahel WR (1982) The product-life factor. Free Trade Reimagined 10(1515/9781400827855):110

Suárez-Eiroa B, Fernández E, Méndez G (2021) Integration of the circular economy paradigm under the just and safe operating space narrative: twelve operational principles based on circularity, sustainability and resilience. J Clean Prod. https://doi.org/10.1016/j.jclepro.2021.129071

Tejaswini MSSR, Pathak P, Gupta DK (2022) Sustainable approach for valorization of solid wastes as a secondary resource through urban mining. J Environ Manage 319:115727. https://doi.org/10.1016/J.JENVMAN.2022.115727

Tuladhar A, Iatridis K, Dimov D (2021) History and evolution of the circular economy and circular economy business models. In: Circular economy and sustainability: management and policy, vol 1. Elsevier, pp 87–106. https://doi.org/10.1016/B978-0-12-819817-9.00031-4

Turner GM (2008) A comparison of the limits to growth with 30 years of reality. Glob Environ Change 18(3). https://doi.org/10.1016/j.gloenvcha.2008.05.001

U.S. Environmental Protection Agency (2023) Use of energy explained Energy use for transportation. https://www.eia.gov/energyexplained/use-of-energy/transportation.php. Accessed 21 August 2025

U.S Enviromental Protection Agency (2025) The solid waste management toolkit for developing countries. https://www.epa.gov/international-cooperation/solid-waste-management-toolkit-developing-countries

Ugalde Hernández O (2021) Historic-Epistemological evolution of the circular economy: towards a new development paradigm? TT. Economía y Sociedad 26(59):83–95. http://www.scielo.sa.cr/pdf/eys/v26n59/2215-3403-eys-26-59-83.pdf

UNEP (2011) Annual report. https://www.unep.org/resources/annual-report/unep-2011-annual-report. Accessed 17 July 2025

UNEP (2014) Assessing global land use: balancing consumption with sustainable supply. https://www.resourcepanel.org/reports/assessing-global-land-use

UNEP (2019) Global resources outlook 2019: summary for policymakers. www.resourcepanel.org/reports/global-resources-outlook-2019.

UNEP (2021) Making peace with nature. a scientific blueprint to tackle the climate, biodiversity and pollution emergencies. Making Peace with Nature. https://doi.org/10.18356/9789280738377

UNEP (2023) Building materials and the climate: constructing a new future. United Nations Environment Programme. https://wedocs.unep.org/xmlui/handle/20.500.11822/43293. Accessed 7 July 2025

UNEP (2024a) Emission Gap Report 2024: No more hot air...please! https://www.unep.org/resources/emissions-gap-report-2024

UNEP (2024b) Global resources outlook 2024: bend the trend—pathways to a liveable planet as resource use spikes. International resource panel

UNESCO (2023) World water development report 2023. UNESCO World Water Assessment Programme. https://www.unesco.org/reports/wwdr/2023/es/download. Accessed 7 July 2025

Usman M, Balsalobre-Lorente D (2022) Environmental concern in the era of industrialization: can financial development, renewable energy and natural resources alleviate some load? Energy Policy 162:112780. https://doi.org/10.1016/J.ENPOL.2022.112780

World Bank (2022) Squaring the circle: policies from Europe's Circular economy transition. https://www.worldbank.org/en/region/eca/publication/squaring-circle-europe-circular-economy-transition

World Bank (2025) Global economic prospects. International Bank for Reconstruction and Development. https://doi.org/10.2307/j.ctt183pb3w.5

Xu B, Chaudhry SS, Li Y (2009) Factors of production: historical theories and new developments. Syst Res Behav Sci 26(2):219–224. https://doi.org/10.1002/SRES.972

Zavos S, Lehtokunnas T, Pyyhtinen O (2024) The (missing) social aspect of the circular economy: a review of social scientific articles. Sustain Earth Rev 7(1). https://doi.org/10.1186/S42055-024-00083-W

Chapter 3
Advancing Circular Models in Practice

Abstract This chapter aims to examine how circular economy models are being operationalized across multiple levels—macro, meso, micro, and nano—through coordinated strategies, governance frameworks, and practical tools. It begins by exploring the role of international organizations and national governments in shaping circular policy agendas, highlighting the importance of strategic alignment and fiscal reform to support systemic transitions. At the meso-level, the chapter analyzes industrial symbiosis, eco-industrial parks, and sector-wide redesigns that facilitate resource sharing and waste reduction. Case studies from Denmark, Finland, and Mexico illustrate how regional collaboration fosters circular innovation. At the micro-level, the chapter investigates community-driven initiatives and small business practices, such as repair cafés and product-service systems, which extend product lifecycles and promote behavioral change. It also introduces the nano-level, where circularity is embedded in product design and material efficiency through tools like the Material Circularity Indicator and Lifecycle Assessment. The chapter emphasizes that circularity is not a one-size-fits-all solution but a dynamic framework requiring context-sensitive implementation. To support this transition, the chapter presents systemic approaches such as industrial ecology and preventive strategies like pollution prevention and cleaner production. These frameworks prioritize source reduction and process redesign, offering alternatives to reactive waste management. Tools including eco-efficiency metrics, environmental audits, and green procurement are discussed as enablers of sustainable consumption. Ultimately, this chapter argues that advancing circular models in practice demands more than technical innovation—it requires ethical commitment, inclusive governance, and cultural transformation. By integrating circular principles across scales and sectors, societies can move toward regenerative systems that align economic activity with ecological integrity and social well-being.

Keywords Industrial ecology · Cleaner production · Pollution prevention · Multilevel governance · Material circularity · Energy recovery · Systemic perspective · Waste valorization

L. Velazquez, *Circular Economy and Sustainable Consumption in the Context of SDG 12*, Strategies for Sustainability, https://doi.org/10.1007/978-3-032-15598-6_3

3.1 Introduction

In today's business landscape, the circular economy is no longer a niche concept—it has become a strategic imperative. Companies are increasingly expected to adopt sustainable practices that not only reduce environmental impact but also drive innovation, resilience, and long-term value creation. Circularity offers a pathway to optimize resource use, minimize waste, and enhance brand reputation in a competitive global market. According to the World Economic Forum (2025), circularity has become a competitive advantage for businesses. Over 75% of companies recognize its importance—up from 40% 3 years ago—and investments in circular initiatives have increased by more than 70%. Despite this growing momentum, there is still no single, unified framework to guide the implementation of circular economy principles (De Angelis 2021).

The previous chapter examined the core aspects of the circular economy, which are essential to understanding and applying its principles in a clear, compelling, and realistic way. To fully operationalize its principles within broader sustainability frameworks, a more sustainable production and consumption system must be grounded in systems thinking and lifecycle assessment. Systems thinking provides a holistic view, understanding reality as a web of interconnected parts, each influencing and being shaped by the others (Hassan and Faggian 2023). Meanwhile, lifecycle assessment encourages a careful examination of the environmental impacts of goods and services, tracking their journey from resource extraction and manufacturing to use and final disposal (Onat et al. 2017).

Systems thinking and lifecycle assessment act as core approaches for developing practical circularity solutions, enabling the integration of diverse knowledge domains and fostering collaboration across several sectors (Entezarian and Mehraeen 2024). Their application helps align technical innovation with environmental responsibility and social equity, guiding the creation of development models that are both holistic and regenerative. This merging of interdisciplinary perspectives from various subsystems allows key stakeholders to overcome siloed visions, expand the influence of decisions, and foster an ethical and regenerative relationship with ecosystems (Iacovidou et al. 2021). However, the integration of diverse domains and actors also introduces significant complexity, requiring alignment across multiple scales and sectors.

3.2 Multi-level Governance and Strategic Alignment for Circularity

Implementing circular economy principles demands coordinated action at macro-, meso-, micro-, and nano-levels (Drofenik et al. 2025), where policies, business models, individual behaviors, and product-service systems must converge to enable systemic transformation. Each level offers unique strategies, frameworks, and tools that reflect the complexity of moving toward more sustainable systems. Understanding how these levels interact is essential for designing fair and practical pathways to circularity. Effective coordination helps bridge the gap between intention and execution, ensuring that circular strategies are adapted to diverse capacities, contexts, and objectives.

At the macro-level, international organizations and governments must take the lead in driving the transition from a linear economic model to a circular economy, fostering systemic change through policy innovation, cross-sector collaboration, and long-term sustainability strategies. While both actors are essential, their roles are distinct yet complementary. According to İbraimi and Hoca (2025), international organizations provide strategic direction, generate evidence-based insights, and create platforms for global cooperation, laying the groundwork for circularity to flourish across borders. Meanwhile, governments, on the other hand, hold the regulatory and fiscal tools necessary to implement these visions at the national level, translating global frameworks into actionable policies and resource allocations.

Building on this differentiated leadership, international organizations have become instrumental in shaping the global transition toward circularity. For instance, the United Nations (UN), through the 2030 Agenda for Sustainable Development, has positioned the circular economy as a key enabler for achieving multiple Sustainable Development Goals (SDGs), particularly those related to responsible consumption and production, climate action, and sustainable cities. The UN promotes systems-based approaches and deep-rooted transformations that redesign production and consumption patterns to stay within planetary boundaries. The Organization for Economic Co-operation and Development (OECD), on its part, is actively supporting countries in developing circular economy strategies by offering targeted policy recommendations and facilitating stakeholder dialogues, which help governments identify priority areas and design effective frameworks for implementing circularity (OECD 2025). Similarly, the World Bank has acknowledged the pressing need to decouple economic growth from resource consumption and advocates for fiscal reforms—such as shifting tax burdens from labor to resource extraction and waste generation—as a means to incentivize more sustainable practices. Moreover, the World Bank calls for scaling up circular business models from niche applications to mainstream economic systems, emphasizing that such a transition is essential for long-term resilience and environmental sustainability (World Bank 2022).

While international organizations provide the strategic vision, global frameworks, and collaborative platforms necessary to guide the transition toward circularity, it is ultimately national governments that operationalize these guidelines. Their role at the macro-level is critical, as they translate international guidance into domestic policy frameworks, infrastructure planning, and systemic governance models that embed circularity into long-term development strategies. This alignment between global direction and national implementation is essential for achieving meaningful progress. For example, the United States Environmental Protection Agency (EPA) has implemented initiatives such as the National Strategy to Prevent Plastic Pollution and the National Strategy for Reducing Food Loss and Waste and Recycling Organics, both of which promote circular principles by minimizing waste and reusing materials (EPA

2025a). China has also emerged as a global leader in circular economy implementation, integrating circular principles into its national development strategies through top-down policy frameworks embedded in its Five-Year Plans (Bleischwitz et al. 2022). The country has focused on transforming traditional industrial parks into eco-industrial parks, promoting industrial symbiosis and closed-loop production systems (Mathews et al. 2018). These efforts have led to measurable improvements in waste reduction and resource efficiency, particularly in manufacturing and urban planning. Furthermore, China's policies emphasize collaboration between large enterprises and small- and medium-sized enterprises (SMEs) to foster innovation and scale sustainable practices across supply chains, despite challenges such as uneven regional implementation and limited SME engagement (Chen 2023; Ogunmakinde 2019). However, the effectiveness of circular economy initiatives remains uneven across regions. Winans et al. (2017) highlight that while circular economy strategies have shown promise—particularly in China's eco-industrial parks, many initiatives suffer from a lack of standardized metrics, inconsistent definitions, and limited empirical evaluation. Similar efforts are underway in countries across Europe, Asia, Latin America, and Africa, where governments are developing national circular economy roadmaps, implementing fiscal reforms, and establishing inter-ministerial commissions to coordinate systemic transitions. According to the United Nations Industrial Development Organization (UNIDO), these types of initiatives reflect a growing global momentum toward circularity, although their scope, governance structures, and effectiveness vary significantly across regions (UNIDO 2024).

At the meso-level, industries and regional actors play a crucial role in closing material loops and maintaining resource value by facilitating circular flows of resources and knowledge through coordinated networks (Ghisellini et al. 2023). This level supports practices such as industrial symbiosis, sector-wide redesign of supply chains, reverse logistics, and modular product design, often enabled by regional policies that foster collaborative innovation (Monteiro and Dutra 2020). These dynamics are frequently observed in eco-industrial parks and municipal projects, where businesses co-locate and share resources to minimize waste and optimize material use (Zhao et al. 2018). A prominent example is the Kalundborg Symbiosis in Denmark, widely recognized as one of the earliest and most successful eco-industrial parks. In Kalundborg, companies exchange energy, water, and byproducts, turning waste from one process into input for another, thereby creating a closed-loop system that reduces environmental impact and enhances economic efficiency (Paché 2024). Similarly, the ECO3 industrial park in the Tampere Region of Finland demonstrates how city-owned development companies can act as intermediaries to align stakeholder interests and catalyze regional circular economy initiatives (European Circular Economy Stakeholder Platform 2025). This park integrates bioeconomy and circular economy principles, enabling collaboration between public and private actors to develop sustainable industrial ecosystems (Nylén et al. 2025).

A similar pattern appears in automotive clusters where major manufacturers collaborate with networks of satellite suppliers following the just-in-time production approach. These closely connected supply chains not only reduce inventory and transportation costs but also support circular practices like remanufacturing components, sharing logistics, and recycling materials locally. The proximity of firms makes it easier to exchange byproducts and reuse materials, while industrial parks provide the infrastructure and policies needed to promote collaboration. In the Mexican context, Pinzón-Castro et al. (2025) highlight how green strategies implemented by manufacturing companies enable the integration of processes such as remanufacturing, eco-design, and material reuse, generating positive impacts on overall sustainable performance. These practices are not only applied within each company but also extend to their relationships with suppliers and industry partners, creating a collaborative environment that promotes circularity. Internationally, Cao et al. (2025) emphasize that the remanufacturing of components, the recycling of critical materials, and the reuse of batteries in electric vehicles are being adopted by multiple stakeholders within automotive clusters, demonstrating that circularity is consolidating as a shared strategy among companies, beyond individual efforts. Building on these premises, McKinsey & Company (2025) forecasts that by 2030, 10% of new vehicles sold will be second hand, driven by the rise of design strategies focused on disassembly, reuse, and recycling.

At the micro-level, individuals, households, communities, and small businesses play a foundational role in advancing circular economy principles through everyday actions and localized innovation (Patil et al. 2023). These actors engage in practices such as repairing, reusing, repurposing, composting, and recycling, which collectively reduce waste generation and extend product lifecycles. One prominent example is the global network of Repair Cafés, which has expanded to over 2,000 locations in 37 countries (Repair Café 2022). These are community-based initiatives where volunteers with technical skills help individuals repair broken household items—such as electronics, clothing, furniture, and bicycles—free of charge. The goal is not only to extend the lifespan of products and reduce waste, but also to empower citizens through knowledge-sharing, promote a culture of care and maintenance, and foster social cohesion. Repair Cafés serve as informal learning environments where participants gain practical skills and awareness about sustainable consumption (Moalem and Mosgaard 2021). SMEs also play a key role in driving circularity at the micro-level. Despite facing challenges such as limited financial and technical resources, many SMEs have adopted innovative circular strategies, including remanufacturing, product-service systems, and closed-loop supply chains (Rizos et al. 2016). For example, SMEs in the manufacturing sector have implemented practices such as reusing metal scraps in production, repairing and upgrading machinery to extend its lifecycle, and recovering heat or water for internal reuse or exchange with nearby firms. In the food industry, SMEs have adopted anaerobic digestion to convert organic waste into biogas or compost, and have optimized packaging to reduce material use and improve recyclability (Saari et al. 2024). A thematic review of circular economy practices among SMEs identified key enablers such as leadership commitment, customer awareness, and access to green technologies, which facilitate the shift to circular models (Kamal and Jasni 2023). Additionally, digital tools and industrial platforms have supported SMEs in tracking material flows, optimizing resource use, and collaborating in local industrial symbiosis networks (Antikainen et al. 2018).

The existence of a nano-level within the circular economy is also acknowledged, consisting of circular products and services that represent the basic unit and fundamental cell of the economic dynamic required for the transition to a circular model. This level encompasses the design and manufacturing stages using circular principles, the maximum extension of product lifespan, and end-of-life recycling (Patil and Langen 2023). At the nano-level, several tools are employed to assess the circularity of products and services, such as the Material Circularity Indicator (MCI), which focuses on resource efficiency around the product (Sekti and Singgih 2022), and Lifecycle Assessment (LCA), which evaluates the environmental impact of a product across all stages of its lifecycle, including raw material extraction (Niero and Kalbar 2019). Nanotechnology is also a key means for promoting a circular economy at the nanoscale, enabling the design of materials and products with functional, reusable, and recyclable properties from their inception (Gilbertson and Vikesland 2022). In the agricultural sector, the integration of biotechnological processes in the manufacturing of nanomaterials has made it possible to transform waste into high-value resources (Teixeira 2025). According to Darwesh et al. (2024), the biological synthesis of zinc oxide (ZnO) nanoparticles promotes resource efficiency by transforming textile waste into multifunctional materials. These nanoparticles confer antimicrobial, ultraviolet protection, and mechanical strength properties to recycled textiles, thus extending their lifespan and expanding their applications. The commendable aspect of this technology lies not only in reducing waste generation but also in avoiding the use of toxic chemical reagents.

Having explored the circular economy across macro-, meso-, micro-, and nano-levels, it becomes clear that achieving systemic circularity requires more than isolated actions—it demands integrated frameworks that guide implementation across sectors and scales. In this context, it's important to recognize that the classification of these levels does not reflect the internal diversity within each. Implementation differs significantly depending on geography, economic strength, and institutional capacity. Countries in the Global North, such as Germany or the Netherlands, typically have the resources and regulatory support to adopt advanced circular strategies (Apoorva and Bhutani 2022). In contrast, those in the Global South, including India or Kenya, often rely on informal networks and community-led initiatives (International Labor Organization 2023). Similarly, multinational corporations have the capacity to overhaul entire supply chains for circularity, but small- and medium-sized businesses frequently face challenges like limited funding, technical expertise, and policy support (Gennari 2023).

To move from principles to practice, it is essential to adopt structured approaches that align environmental goals with industrial processes and market dynamics. The following section, Circular Economy from a Systemic Perspective, introduces the framework of industrial ecology, which emphasizes material and energy flows within interconnected systems, promoting symbiotic relationships among industries. Complementing this, the section Circular Economy in Business and Market Dynamics explores preventive approaches such as pollution prevention and cleaner production, which focus on minimizing waste and resource use at the source through innovation, efficiency, and redesign. Together, these frameworks provide a foundation for embedding circularity into both systemic structures and everyday business operations.

3.3 Circular Economy from a Systemic Perspective

Industrial Ecology emerged as a new way of understanding production systems, shifting the perspective from viewing industry as a simple linear process of production and disposal to seeing it as an interconnected ecosystem, much like those found in nature (Saavedra et al. 2018). This perspective was strongly influenced by the work of Frosch and Gallopoulos (1989) who proposed that one company's waste could become a valuable resource for another, mirroring the principles of biological ecosystems. Their ideas are closely aligned with the concept of industrial metabolism, as developed by Robert U. Ayres, which analyzed the flows of materials and energy through industrial systems, laying a foundational framework for the emerging field (Ayres 1998). In an industrial ecological system, materials and energy can be utilized efficiently, resulting in reduced pollution and the effective reuse of every waste, scrap, and byproduct (Wang et al. 2014). However, the authors warned that creating a fully functional industrial ecosystem like the one described is a challenging goal because of constraints such as technological limitations, economic pressures, and a lack of widespread understanding, although the concept itself was valuable. The authors also emphasized the importance of spreading the principles behind industrial ecology and systems optimization, which was seen as a critical step toward rethinking how products are made and how industries operate.

Building on these pioneering insights, Allenby (1992) proposed three models within the framework of industrial ecology that illustrate the shift from a linear production system to one where circularity is nearly complete. The Type I model shows a linear and simplified view of industrial processes, assuming that natural resources and environmental sinks are always available and unlimited. This industrial model of production has been criticized for the lack, or near absence, of restrictions on the extraction, transformation, use, and disposal of resources, mainly materials but also energy. Circularity appears with the Type II model as raw materials become scarcer, leading to increased efforts to reduce waste and environmental damage through reuse and recycling. This represents a moderate step toward closing resource loops and respecting ecological limits, but it does not achieve full circularity. Finally, the Type III model embodies the ideal state within industrial ecology. In this framework, all materials circulate continuously without generating waste, scrap, or byproducts; therefore, there is no need to extract new inputs. In other words, the resource cycle is completely closed. However, although it is closed in terms of matter, it remains open in terms of energy, as it requires a constant flow of external energy, such as solar radiation, to sustain active processes.

More than just theoretical ideas, model type II and model type III provided practical insights for scholars to examine how industrial systems might adapt to ecological and resource pressures. These essential insights became part of a larger process that shaped the emerging field of industrial ecology. Among several key milestones, the U.S.-Japan Workshop on Industrial Ecology, held in March 1993 in the United States, marked a pivotal moment in the development of industrial ecology. Although it did not aim for immediate consensus, the meeting planted ideas that would later evolve into theoretical frameworks, applied methodologies, and international cooperation networks dedicated to fostering the field (Richards and Fullerton 1994).

Within just a few years of its emergence, industrial ecology began to face critical reflection regarding its conceptual and disciplinary foundations. Rather than remaining a metaphor derived from natural ecosystems, it was increasingly examined as a potential academic field with its own methodologies and institutional identity. Several key publications began to clarify the principles of Industrial Ecology. Books like Industrial Ecology, Graedel and Allenby (1995) deepened the discussion about industrial metabolism, circularity, and industrial ecosystems, and offered practical tools to put these concepts into action. Gradually, these contributions turned industrial ecology from a promising metaphor into a well-established field discipline, capable of inspiring real change across sectors. Despite the development of analytical tools, questions persisted about its status as discipline.

Today, the metaphor of industrial systems that mimic the circular and resilient functions of nature has evolved into tangible initiatives that demonstrate the practical viability of industrial ecology. As global economies face increasing pressure to decouple growth from environmental harm, industrial ecology offers a systems-based approach that integrates material, energy, and information flows across industrial networks and policy landscapes (Ahmadov et al. 2025; Cagno et al. 2023). It not only supports the design of closed-loop systems and symbiotic industrial clusters but also guides national and regional circular economy strategies (Bouillon et al. 2025; Urbinati et al. 2021). Industrial ecology emphasizes optimizing the entire materials cycle—from raw materials to final disposal—while balancing economic competitiveness with environmental responsibility (Yuan and Bi 2006). Its contributions to the circular economy encompass conceptual, technical, and policy areas (Saavedra et al. 2018). While the circular economy defines goals such as resource efficiency, waste minimization, and regenerative systems, industrial ecology provides the analytical methods and systemic approaches needed to achieve those goals effectively (Cecchin et al. 2021).

As industrial ecology matured, its scope expanded beyond technical optimization and resource efficiency. Industrial ecology evolved into more than a technical framework—it became a transformative approach that integrates ethical, cultural, and systemic dimensions. By challenging dominant paradigms, it offers a pathway toward a more sustainable and equitable future. However, as Ehrenfeld (2004) argued, its consolidation as a discipline required not only methodological rigor and institutional support, but also a broader normative vision capable of guiding its development beyond technical boundaries.

3.4 Circular Economy in Business and Market Dynamics

Circular economy tools redefine business practices and market dynamics by embedding preventive strategies into key areas such as production processes, product design, and supply chain management (Kazakova and Lee 2022). For instance, Dennison et al. (2024) note how design for circularity increases the durability, reusability, and recyclability of products. Moreover, the authors provide evidence from the literature about regenerative design, which is aimed at restoring and revitalizing natural systems. This transformation is further supported by evolving regulatory frameworks that encourage companies to adopt resource-efficient models, reduce waste, and extend product lifecycles, thereby influencing how markets function and how value is generated (Hondroyiannis et al. 2024). In response to this structural and normative evolution, more businesses are embracing circular economy principles to enhance resilience, reduce operational costs, respond to growing consumer and regulatory expectations for sustainability, significantly mitigate safety risks, and unlock new economic opportunities by decoupling profitability from resource consumption (Constance Obiuto et al. 2024; Munonye 2025; Rodríguez-Espíndola et al. 2022). These principles include eliminating waste and pollution through thoughtful design, keeping products and materials in use to maximize their value, and regenerating natural systems to restore ecological balance (Velenturf and Purnell 2021).

The evolution of waste management from reactive disposal to proactive prevention reflects a broader shift toward circularity. At the core of many circular economy strategies lies a preventive approach that aims to eliminate environmental harm at the source instead of relying on reactive, end-of-pipe solutions—such as filters or treatment systems—these strategies promote systemic innovation and redesign, ensuring sustainability is built into processes from the start (Omer et al. 2025). While these reactive measures can reduce environmental damage, they often come with additional costs and fail to target the root causes of pollution. Furthermore, they can negatively affect profitability because they do not add value to the product or process but instead increase operational expenses. If not adequately managed, end-of-pipe solutions may also cause unintended transfer of pollutants from one environmental medium to another (Sun et al. 2023). For example, air filters might effectively capture airborne contaminants, but without proper disposal or treatment, the waste collected could contaminate water or soil—shifting the environmental burden rather than solving it. By addressing impacts before they happen, preventive strategies support material circularity and lessen dependence on downstream waste treatment.

Within the framework of the circular economy, waste management assumes a strategic role by guiding decisions toward practices that optimize resource use and minimize environmental impact. Central to this approach is the waste management hierarchy, a conceptual model that ranks waste handling methods from most to least preferred. Commonly depicted as an inverted pyramid, the hierarchy begins with source reduction, followed by reuse, recycling, recovery, treatment, and finally disposal. Its formalization in policy began with the European Union's Waste Framework Directive 2008/98/EC, which established a five-tier priority order for waste operations (Zhang et al. 2022). However, the hierarchy is not static; it has evolved to reflect regional priorities and sector-specific needs. For example, in some frameworks, the recovery stage is subdivided into energy recovery and material recovery, depending on the technological and regulatory context (Awino and Apitz 2024). In simplified models, the treatment phase is occasionally omitted, which can obscure the environmental implications of intermediate processes (Apaydın 2025).

According to the 2019 EPA's TRI National Analysis, the most common source reduction activities in U.S. industry were good operating practices (40%), process modifications (21%), and spill and leak prevention (10%). Good source reduction practices often involve improved maintenance, employee training, and procedural adjustments, while process modifications focus on changes in equipment or technology to reduce material use. Spill and leak prevention centers on containment and monitoring systems to prevent unintended releases (EPA 2021). These commonly adopted practices reflect the broader philosophy of prevention, which has evolved over time into more structured approaches. Although the upper levels of the waste management hierarchy align with circular economy principles by promoting dematerialization and reduced resource extraction, it is important to recognize that these outcomes are neither automatic nor guaranteed. As Bandh et al. (2023) and Nanda and Berruti (2021) caution, following the hierarchy or adopting resource productivity-oriented approaches does not always lead to an absolute decrease in material flow. Given that full circularity is not always achievable in practice, it becomes essential to enhance performance even within the lower tiers of the waste management hierarchy. Recovery and disposal, while generally considered less desirable due to their limited resource retention and higher environmental impact, remain necessary components in specific contexts (Bandh et al. 2023; Nanda and Berruti 2021). To address this, the United Nations Environment Programme (UNEP) developed the Integrated Waste Management Scoreboard, a tool designed to evaluate and improve waste management practices across all stages, including final disposal (UNEP 2006). In this phase, the scoreboard supports municipalities and institutions in assessing landfill operations, environmental safeguards, and long-term sustainability strategies, thereby promoting accountability and continuous improvement even in the least preferred tier. To further support the operationalization of waste management practices across the hierarchy, the United Nations Environment Programme's Global Waste Management Outlook (2024) emphasizes that advancing toward the top of the hierarchy—where prevention stands as a cornerstone—requires coordinated action among governments, businesses, industry, and consumers. Each actor contributes through policymaking, product design, technological innovation, and responsible consumption, enabling systemic change and reducing material throughput (UNEP 2024a).

Among the most influential preventive frameworks that operationalized systemic practices aligned with the waste management hierarchy are pollution prevention and cleaner production, both of which laid the necessary groundwork for circularity by promoting strategies to reduce pollution at the source. They became key methods for environmental management in the late 1980s and early 1990s. Although both share a preventive philosophy aimed at reducing pollution at its source, they emerged from distinct institutional and geographic contexts. Pollution prevention was formally established in the United States through the Pollution Prevention Act of 1990, which designated source reduction as the primary national environmental policy under the U.S. Environmental Protection Agency (EPA 1990). In contrast, cleaner production was initially promoted through international and European policy frameworks, particularly in countries with strong manufacturing sectors, reflecting a broader global commitment to integrating preventive strategies into industrial development (Van Berkel 2011).

Pollution prevention encourages organizations to rethink how they design, operate, and manage their processes by integrating sustainability into core decision-making. As defined by the EPA, it encompasses practices that reduce or eliminate pollution at its source—before it is created—marking a shift from reactive to proactive environmental strategies. In alignment with this perspective, the United Nations Environment Programme defines pollution prevention as reducing or eliminating waste and pollutants at the source, emphasizing proactive strategies over reactive cleanup (UNEP 2006). Similarly, the Organization for Economic Co-operation and Development promotes integrated pollution prevention strategies that support circular economy principles and long-term climate goals (OECD 2021), while the United Nations Economic Commission for Europe (UNECE) also advocates pollution prevention through cross-border cooperation, highlighting the need for coordinated policies to tackle air pollution across nations (UNECE 2024). The long-term success of pollution prevention initiatives in the private sector demonstrates their ability to deliver both environmental and economic advantages. According to the EPA (2024), pollution prevention strategies often result in significant financial savings by reducing waste, easing regulatory burdens, and improving operational efficiency. Scholars contend that pollution prevention investments offer high returns by lowering liability risks, enhancing brand reputation, and fostering innovation (Reichel 2023), while the Organization for Economic Co-operation and Development emphasizes that integrating pollution prevention into industrial policy supports long-term competitiveness and resource efficiency (OECD 2021).

One of the most prominent and long-standing pollution prevention initiatives is 3 M's Pollution Prevention Pays program, launched in 1975, aimed to eliminate pollution at its source through strategies such as product reformulation, process modification, and equipment redesign, moving away from traditional end-of-pipe solutions. Recognized by the EPA as a benchmark for integrating pollution prevention into core business operations, the initiative shows how continuous source reduction efforts can produce tangible results. For example, the program saved the company over $1 billion while preventing the release of billions of pounds of pollutants, demonstrating that environmental responsibility can align with economic performance (EPA 2025b).

Following 3 M's example, other businesses adopted pollution prevention as a key part of their waste reduction strategies. From the mid-1980s through the early 2000s, pollution prevention gained momentum across North America, with numerous companies reengineering their operations to minimize waste at its source. According to the Commission for Environmental Cooperation of North America (2004), in the United States, Canada, and Mexico implemented cleaner production practices, process modifications, and material substitutions to reduce environmental impacts. Many of these efforts were supported by government and industry-led programs that allocated resources to demonstration projects, showcasing the technical and economic feasibility of pollution prevention in diverse industrial sectors. Moreover, the Organization for Economic Co-operation and Development report on best available techniques for industrial pollution prevention and control presents an updated overview of how different countries have adopted integrated approaches to improve the environmental performance of the productive sector (OECD 2024). Rather than replacing traditional emission control methods, these techniques are conceived as a complementary strategy that promotes prevention at source through process redesign, the substitution of hazardous inputs, and the adoption of clean technologies. Countries such as Germany, Japan, the United States, India, China, New Zealand, and members of the European Union have incorporated several of these techniques into their regulatory frameworks, linking them to environmental permits and sectoral standards. This convergence between prevention and control has significantly reduced emissions to air, water, and soil, while boosting technological innovation and industrial competitiveness.

Similarly, cleaner production represents another influential framework whose principles are tightly aligned with the circular economy's goals of resource circularity, waste minimization, and product longevity (Almeida et al. 2017). Introduced by the United Nations Environment Programme in 1989 as a global strategy for sustainable industrial development, cleaner production promotes the continuous application of preventive practices in processes, products, and services to minimize environmental impacts at the source (UNEP 2012). While it shares core concepts with most pollution prevention strategies, cleaner production is characterized by its direct implementation within industrial processes, providing a specialized way to incorporate circularity into business operations (de Mello Santos et al. 2021). Over time, it has evolved into a distinctly technical and engineering-oriented approach, emphasizing process redesign, optimizing input–output ratios, reducing entropy in material flows, substituting hazardous materials, and integrating cleaner technologies into the production system (Gomes da Silva and Gouveia 2020).

Over the past several decades, Pollution Prevention and Cleaner Production have not only evolved as complementary frameworks, but have also been increasingly applied in tandem across industrial and policy contexts. Their shared emphasis on source reduction and proactive environmental management has led to overlapping strategies and integrated approaches, reinforcing their relevance within circular economy initiatives and sustainable development agendas.

3.5 Tools for Sustainable Consumption in Industrial Ecology, Cleaner Production, and Related Fields. Pollution Prevention

Industrial Ecology and Pollution Prevention and Cleaner Production are mutually reinforcing approaches within the broader framework of corporate sustainability. Although they differ in scale and emphasis, both seek to minimize environmental impact through systemic transformation and resource efficiency, ultimately advancing sustainable consumption.

Industrial ecology typically operates at macro- or meso-levels, utilizing tools such as Material Flow Analysis and Industrial Symbiosis Mapping to facilitate the exchange of energy, water, and materials among firms—particularly within eco-industrial parks (Chertow 2007; Kalmykova et al. 2016). These instruments support the design of circular systems that reduce embedded environmental costs and foster sustainable consumption across interconnected networks. Conversely, pollution prevention and cleaner production focus on micro-level interventions, applying methodologies like lifecycle assessment, environmental management systems, and cleaner production audits to optimize internal operations and mitigate emissions at their source (Lozano 2015; Wang et al. 2019). Preventive strategies are further operationalized through tools such as eco-efficiency metrics, resource intensity indicators, and waste minimization checklists, which guide upstream decision-making to avoid environmental burdens before they materialize (De Araujo and De Oliveira 2012).

At the product development stage, design for environment integrates environmental criteria into design processes, promoting modularity, recyclability, and reduced material intensity—thereby shaping consumer behavior and reinforcing sustainable consumption from inception (Bovea and Pérez-Belis 2012). Complementary to this, product-service system design encourages access-based consumption models that lower material throughput while maintaining user satisfaction (Tukker 2015). Advanced analytical tools such as exergy analysis and emergy accounting provide deeper insights into the quality and origin of energy and material flows, enabling prioritization of actions that respect ecological boundaries (Brown and Ulgiati 2016). Meanwhile, chemical footprinting and green supply chain management assess and mitigate toxic impacts throughout the value chain (Sala and Goralczyk 2013).

Additional techniques include pinch analysis for optimizing energy and resource integration (Klemeš et al. 2013), hazardous substance substitution frameworks like GreenScreen for Safer Chemicals (Heine and Franjevic 2013), and real-time monitoring systems that detect inefficiencies and emissions before they escalate (da Costa et al. 2024). Energy and water audits also play a critical role by identifying conservation opportunities aligned with responsible resource use (Andrews and Sturm 2016; Silveira and Mata-Lima 2020; Wrage 2021). Finally, these approaches are complemented by initiatives such as eco-labeling, green procurement, and consumer education, which stimulate market demand for more sustainable products and services (Wang et al. 2019).

Together, these tools form a multi-level architecture that enables a more comprehensive strategy for advancing circular economy goals. By leveraging the strengths of both systemic and operational frameworks, organizations can develop interventions that not only optimize resource flows across industrial networks but also improve performance within individual enterprises. This synergy supports the shift toward more resilient and regenerative production systems, where the emphasis on preventing, eliminating, or minimizing waste and emissions at their source enhances both operational efficiency and environmental performance.

3.6 Recycling: A Strategic Enabler for Sustainable Consumption

Recycling plays a key role in promoting sustainable consumption by enabling the recovery and reuse of materials that would otherwise be discarded. Although it is not classified as a source reduction strategy in the waste management hierarchy, it significantly helps reduce the environmental impacts of consumer habits and production processes.

By decreasing the demand for virgin resources and lowering the energy required for manufacturing, recycling encourages more responsible resource use and helps reduce the ecological footprint of consumption (Jiang and Bateer 2025). According to the United Nations Environment Program, recycling and other circular economy strategies could generate a net global economic benefit of over USD 108 billion annually by 2050, while significantly decreasing pollution and health-related costs (UNEP 2024a). From a consumption perspective, recycling encourages individuals and communities to reflect on the value of materials and the consequences of disposal. It fosters behavioral change by promoting awareness of waste generation and by reinforcing habits such as sorting, reusing, and participating in local recycling programs. As Oberoi et al. (2021) emphasize, recycling contributes to sustainable development not only environmentally but also socially and economically, by creating employment opportunities and supporting local circular initiatives. The World Bank estimates that global waste generation will rise from 2.01 billion tons in 2016 to 3.40 billion tons by 2050, underscoring the urgency of scaling up recycling and waste reduction efforts (World Bank 2018).

Despite its advantages, recycling is not a panacea. Its effectiveness hinges on the availability of infrastructure, active consumer participation, and favorable market conditions—all of which vary significantly across regions. Inadequate systems can lead to inefficiencies, contamination, and low recovery rates. Moreover, certain recycling technologies, especially those involving chemical processes, have sparked concerns due to their high energy demands and potential environmental trade-offs (Kang et al. 2023). Crucially, recycling does not tackle upstream challenges such as overproduction, planned obsolescence, or unsustainable product design. Without addressing these root causes, recycling risks becoming a reactive measure that delays—but does not prevent—environmental degradation.

Nevertheless, when integrated into broader strategies that include eco-design, extended producer responsibility, and sustainable consumption education, recycling becomes a valuable tool for closing material loops. For instance, the OECD emphasizes that recycling must be embedded within broader resource efficiency and circular economy policies to ensure systemic change, rather than being treated as a standalone solution (OECD 2023). The United Nations Environment Program has likewise highlighted that extended producer responsibility not only shifts the burden of post-consumer waste management from municipalities to producers but also stimulates product and packaging redesign to reduce waste generation and environmental impacts throughout the product lifecycle (UNEP 2024b). Furthermore, over 150 organizations, including more than 100 leading businesses, have publicly recognized extended producer responsibility as the only proven mechanism for scaling up recycling systems and ensuring adequate funding for collection, sorting, and processing infrastructure (Ellen McArthur Foundation 2021). Recycling provides a practical entry point for consumers to engage in circular practices, supporting the transition toward more sustainable lifestyles. Again, while not strictly preventive, its role in reducing resource extraction and promoting material consciousness makes it an essential component of sustainable consumption frameworks.

In parallel with recycling, reuse plays a critical role in advancing sustainable consumption by extending the functional life of products and reducing the demand for virgin materials (Psarommatis et al. 2025). Unlike recycling, which often requires energy-intensive processing, reuse strategies preserve the embedded energy and labor of products, making them inherently more resource-efficient. Various industrial sectors actively promote reuse practices, including the textile industry through second-hand clothing markets and take-back programs, the electronics sector via refurbishment and resale of devices, and the construction industry through the recovery and repurposing of building materials (Psarommatis and Azamfirei 2024). These practices not only reduce waste generation but also foster new business models such as product-as-a-service and reverse logistics (Shahzabeen et al. 2023). Moreover, reuse encourages behavioral shifts among consumers by emphasizing durability, repairability, and shared use over disposability. When supported by policies that incentivize product longevity and infrastructure for collection and redistribution, reuse becomes a powerful complement to recycling in closing material loops and promoting more responsible consumption patterns (Cialani et al. 2025; Shahzabeen et al. 2023).

3.7 Conclusion

The circular economy is not a singular solution but a multifaceted framework that requires coordinated action across scales, sectors, and disciplines. This chapter has shown that circularity must be understood as a systemic transformation—one that involves international policy, industrial collaboration, community engagement, and product-level innovation. From macro-level governance to nano-level design, each layer contributes uniquely to the transition, yet none can succeed in isolation.

The integration of preventive strategies, such as pollution prevention and cleaner production, reinforces the importance of addressing environmental impacts at their source. These approaches demonstrate that circularity is not only about managing waste but about rethinking how we produce, consume, and design systems from the outset. Tools like lifecycle assessment, material flow analysis, and eco-design provide the analytical foundation to guide this shift.

Ultimately, advancing circular models in practice requires more than technical solutions—it demands ethical commitment, cultural change, and inclusive participation. The diversity of actors and contexts explored in this chapter reflects both the complexity and the opportunity of circularity. By embracing collaboration, innovation, and systemic thinking, societies can move closer to a regenerative model that aligns economic activity with ecological integrity and social well-being.

References

Ahmadov T, Durst S, Gerstlberger W (2025) Unveiling success factors for implementing and sustaining circular economy practices in small and medium-sized firms: multi-level perspective. Bottom Line 38(1):71–98. https://doi.org/10.1108/BL-12-2023-0320

Allenby BR (1992) Industrial ecology: the materials scientist in an environmentally constrained world. MRS Bull 17(3):46–51. https://doi.org/10.1557/S0883769400040859/METRICS

Almeida CMVB, Agostinho F, Huisingh D, Giannetti BF (2017) Cleaner production towards a sustainable transition. J Clean Prod 142:1–7. https://doi.org/10.1016/J.JCLEPRO.2016.10.094

Andrews L, Sturm R (2016) Water audits in the United States: challenges, successes, and opportunities. Am Water Works Assoc 108(2):24–29. https://www.jstor.org/stable/jamewatworass.108.2.24

Antikainen M, Uusitalo T, Kivikytö-Reponen P (2018) Digitalisation as an enabler of circular economy. Procedia CIRP 73(March):45–49. https://doi.org/10.1016/j.procir.2018.04.027

Apaydın Ö (2025) Analysis of sustainable municipal solid waste management alternatives based on source separation using the analytic hierarchy process. Sustainability (Switzerland) 17(9). https://doi.org/10.3390/SU17093868

Apoorva A, Bhutani A (2022) Estudio comparado d legislación internacional sobre Economía Circular. https://www.undp.org/sites/g/files/zskgke326/files/2022-05/undp-cr-estudio-comparado-legislacion-economia-circular.pdf

Awino FB, Apitz SE (2024) Solid waste management in the context of the waste hierarchy and circular economy frameworks: an international critical review. Integr Environ Assess Manag 20(1):9–35. https://doi.org/10.1002/IEAM.4774

Ayres RU (1998) Industrial metabolism: work in progress. 195–228. https://doi.org/10.1007/978-94-017-3511-7_10

Bandh SA, Malla FA, Wani SA, Hoang AT (2023) Waste management and circular economy. Waste Manag Circul Econ 1–17. https://doi.org/10.1007/978-3-031-42426-7_1

Bleischwitz R, Yang M, Huang B, Xu X, Zhou J, McDowall W et al (2022) The circular economy in China: achievements, challenges and potential implications for decarbonisation. Resour Conserv Recycl 183:106350. https://doi.org/10.1016/J.RESCONREC.2022.106350

Bouillon M, Gelsomino LM, de Goeij C (2025) Industrial symbiosis, a key strategy for building resilient and sustainable supply chains 50–56. https://doi.org/10.1007/978-3-031-92082-0_6

Bovea MD, Pérez-Belis V (2012) A taxonomy of ecodesign tools for integrating environmental requirements into the product design process. J Clean Prod 20(1):61–71. https://doi.org/10.1016/J.JCLEPRO.2011.07.012

Brown MT, Ulgiati S (2016) Emergy assessment of global renewable sources. Ecol Model 339:148–156. https://doi.org/10.1016/J.ECOLMODEL.2016.03.010

Cagno E, Negri M, Neri A, Giambone M (2023) One framework to rule them all: an integrated, multi-level and scalable performance measurement framework of sustainability, circular economy and industrial symbiosis. Sustain Product Consump 35:55–71. https://doi.org/10.1016/J.SPC.2022.10.016

Cao Y, Shen B, Tana S (2025) Enhancing ESG practices in supply chain operations through information disclosure: the "ETHICS" framework. J Bus Res 201:115668. https://doi.org/10.1016/J.JBUSRES.2025.115668

Cecchin A, Salomone R, Deutz P, Raggi A, Cutaia L (2021) What Is in a name? the rising star of the circular economy as a resource-related concept for sustainable development. Circul Econ Sustain 1(1):83–97. https://doi.org/10.1007/s43615-021-00021-4

Chen R (2023) China's Circular economy policies: review and reflection. Circul Innov Lab (4):1–25. https://circulareconomy.europa.eu/platform/sites/default/files/2023-08/China'sCircularEconomyPolicies_ReviewandReflection.pdf

Chertow MR (2007) "Uncovering" industrial symbiosis. J Ind Ecol 11(1):11–30. https://doi.org/10.1162/JIEC.2007.1110

Cialani C, Mortazavi R, Sarcinella F (2025) Material footprint and circular economy for a sustainable consumption and production pattern. Circul Econ Sustain 1–20. https://doi.org/10.1007/S43615-025-00505-7/TABLES/5

Commission for Environmental Cooperation of North America (2004) Moving forward with pollution prevention in north america: a progress update. https://www.cec.org/files/documents/publications/2143-moving-forward-pollution-prevention-in-north-america-en.pdf

Constance Obiuto N, Ninduwezuor-Ehiobu N, Chigozie Ani E, Andrew Olu-lawal K, David Ugwuanyi E (2024) Implementing circular economy principles to enhance safety and environmental sustainability in manufacturing. Int J Adv Multidiscipl Res Stud 4(2):22–29. https://doi.org/10.62225/2583049x.2024.4.2.2432

da Costa TP, da Costa DMB, Murphy F (2024) A systematic review of real-time data monitoring and its potential application to support dynamic life cycle inventories. Environ Impact Assess Rev 105:107416. https://doi.org/10.1016/J.EIAR.2024.107416

Darwesh OM, Matter IA, Al-Balakocy NG, Abo-Alkasem MI (2024) Circular economy reinforcement through molecular fabrication of textile wastes with microbial synthesized ZnO nanoparticles to have multifunctional properties. Sci Rep 14(1). https://doi.org/10.1038/S41598-024-66430-1

De Angelis R (2021) Circular economy and paradox theory: a business model perspective. J Clean Prod 285:124823. https://doi.org/10.1016/J.JCLEPRO.2020.124823

De Araujo JB, De Oliveira JFG (2012) Towards a balanced scoreboard for assessing manufacturing processes sustainability. Int J Bus Perform Manag 13(2):198–221. https://doi.org/10.1504/IJBPM.2012.046201;WGROUP:STRING:PUBLICATION

de Mello Santos VH, Campos TLR, Espuny M, de Oliveira OJ (2021) Towards a green industry through cleaner production development. Environ Sci Pollut Res 29(1):349–370. https://doi.org/10.1007/S11356-021-16615-2

Dennison MS, Kumar MB, Jebabalan SK (2024) Realization of circular economy principles in manufacturing: obstacles, advancements, and routes to achieve a sustainable industry transformation. Discov Sustain 5(1). https://doi.org/10.1007/S43621-024-00689-2

Drofenik J, Seljak T, Novak Pintarič Z (2025) A multi-level approach to circular economy progress: linking national targets with corporate implementation. J Clean Prod 493:144902. https://doi.org/10.1016/J.JCLEPRO.2025.144902

Ehrenfeld J (2004) Industrial ecology: a new field or only a metaphor? J Clean Prod 12(8–10):825–831. https://doi.org/10.1016/J.JCLEPRO.2004.02.003

Ellen MacArthur Foundation (2021) Extended producer responsibility position paper
Entezarian N, Mehraeen M (2024) Systems thinking in the circular economy: an integrative literature review. J Syst Think Pract 3(3):68–90. https://doi.org/10.22067/JSTINP.2024.86123.1086
EPA (1990) Pollution prevention Act of 1990. Pollution Prevention (P2). https://www.epa.gov/p2/pollution-prevention-act-1990. Accessed 6 August 2025
EPA (2021) Pollution prevention and waste management. TRI National Analysis 2019. www.epa.gov/trinationalanalysis/. Accessed 19 September 2025
EPA (2024) Learn about pollution prevention. Pollution Prevention (P2). https://www.epa.gov/p2/learn-about-pollution-prevention. Accessed 6 August 2025
EPA (2025a) National strategy to prevent plastic pollution. Circular Economy. https://www.epa.gov/circulareconomy/national-strategy-prevent-plastic-pollution. Accessed 5 August 2025
EPA (2025b) From "pollution prevention pays" to sustainability at 3m. 3M-Lean Six Sigma and Sustainability. https://www.epa.gov/sustainability/3m-lean-six-sigma-and-sustainability. Accessed 7 August 2025
European Circular Economy Stakeholder Platform (2025) From waste water to biofuels: ECO3 bio- and circular economy business area. Good Practices. https://circulareconomy.europa.eu/platform/en/good-practices/waste-water-biofuels-eco3-bio-and-circular-economy-business-area. Accessed 6 August 2025
Frosch RA, Gallopoulos NE (1989) Strategies for manufacturing. Sci Am 261(2):144–153
Gennari F (2023) The transition towards a circular economy: a framework for SMEs. J Manag Govern. Springer US. https://doi.org/10.1007/s10997-022-09653-6
Ghisellini P, Patil RA, Patil RA, Ramakrishna S (2023) Circularity at meso level: a sector perspective. In: Circularity assessment: macro to nano. Springer Nature, pp 57–74. https://doi.org/10.1007/978-981-19-9700-6_5
Gilbertson LM, Vikesland PJ (2022) Inspiring a nanocircular economy. Environ Sci Nano 9(3):839–840. https://doi.org/10.1039/D2EN90005J
Gomes da Silva FJ, Gouveia RM (2020) Cleaner production. https://doi.org/10.1007/978-3-030-23165-1
Graedel TE, Allenby BR (1995) Industrial ecology. Prentice Hall College Div. https://archive.org/details/industrialecolog0000grae_j8q9. Accessed 17 July 2025
Hassan H, Faggian R (2023) System thinking approaches for circular economy: enabling inclusive, synergistic, and eco-effective pathways for sustainable development. Front Sustain 4. https://doi.org/10.3389/FRSUS.2023.1267282/FULL
Heine LG, Franjevic SA (2013) Chemical hazard assessment and the greenscreen™ for safer chemicals. Chem Alternat Assess 129–156. https://doi.org/10.1039/9781849737234-00129
Hondroyiannis G, Sardianou E, Nikou V, Evangelinos K, Nikolaou I (2024) Circular economy and macroeconomic performance: evidence across 28 European countries. Ecol Econ 215. https://doi.org/10.1016/j.ecolecon.2023.108002
Iacovidou E, Hahladakis JN, Purnell P (2021) A systems thinking approach to understanding the challenges of achieving the circular economy. Environ Sci Pollut Res 28(19):24785–24806. https://doi.org/10.1007/S11356-020-11725-9
İbraimi E, Hoca E (2025) Shifting authority and responsibilities, the role of international organizations in the evolving international system. J Peacebuild Transcult Commun 5(9–10):39–45. https://doi.org/10.62792/UT.FREEDOM.V5.I9-10.P2895
International Labor Organization (2023) Global South circular economy could generate millions of job opportunities. Circular Economy. https://www.ilo.org/es/resource/news/la-economía-circular-del-sur-global-podría-generar-millones-de. Accessed 6 August 2025
Jiang X, Bateer B (2025) A systematic review of plastic recycling: technology, environmental impact and economic evaluation. Waste Manag Res https://doi.org/10.1177/0734242X241310658-FIG7.JPG
Kalmykova Y, Rosado L, Patrício J (2016) Resource consumption drivers and pathways to reduction: economy, policy and lifestyle impact on material flows at the national and urban scale. J Clean Prod 132:70–80. https://doi.org/10.1016/J.JCLEPRO.2015.02.027

Kamal LN, Jasni NS (2023) A thematic review on circular economy practices among smes. In: Proceedings of the international conference in technology, humanities and management (ICTHM 2023), 12–13 June, 2023, Istanbul, Turkey, 131, pp 1096–1110. https://doi.org/10.15405/epsbs.2023.11.89

Kang Z, Huang Z, Peng Q, Shi Z, Xiao H, Yin R et al (2023) Recycling technologies, policies, prospects, and challenges for spent batteries. iScience 26(11):108072. https://doi.org/10.1016/J.ISCI.2023.108072

Kazakova E, Lee J (2022) Sustainable manufacturing for a circular economy. Sustainability (Switzerland) 14(24). https://doi.org/10.3390/SU142417010

Klemeš JJ, Varbanov PS, Kravanja Z (2013) Recent developments in process integration. Chem Eng Res Design 91(10):2037–2053. https://doi.org/10.1016/J.CHERD.2013.08.019

Lozano R (2015) A holistic perspective on corporate sustainability drivers. Corp Soc Responsib Environ Manag 22(1):32–44. https://doi.org/10.1002/CSR.1325;WGROUP:STRING:PUBLICATION

Mathews JA, Tan H, Hu MC (2018) Moving to a circular economy in China: transforming industrial parks into eco-industrial parks. Calif Manage Rev 60(3):157–181. https://doi.org/10.1177/0008125617752692

McKinsey & Company (2025) Industria automotriz europea: Lo que necesita para recuperar la competitividad | McKinsey. https://www.mckinsey.com/featured-insights/destacados/industria-automotriz-europea-lo-que-necesita-para-recuperar-la-competitividad/es. Accessed 6 August 2025

Moalem RM, Mosgaard MA (2021) A critical review of the role of repair cafés in a sustainable circular transition. Sustainability 13(22):12351. https://doi.org/10.3390/SU132212351

Monteiro MAP, Dutra CJ (2020) Circular economy meso-level planning : an approach with 'distributed economies.' In: The routledge handbook of waste, resources and the circular economy, pp 153–161. https://doi.org/10.4324/9780429346347-17

Munonye WC (2025) Integrating circular economy principles in business strategies: a policy-driven approach. Circul Econ Sustain 5(3):1887–1896. https://doi.org/10.1007/S43615-025-00523-5/FIGURES/2

Nanda S, Berruti F (2021) Municipal solid waste management and landfilling technologies: a review. Environ Chem Lett 19(2):1433–1456. https://doi.org/10.1007/S10311-020-01100-Y/METRICS

Niero M, Kalbar PP (2019) Coupling material circularity indicators and life cycle based indicators: a proposal to advance the assessment of circular economy strategies at the product level. Resour Conserv Recycl 140:305–312. https://doi.org/10.1016/J.RESCONREC.2018.10.002

Nylén EJ, Anttiroiko AV, Tiensuu A (2025) The role of eco-industrial parks in promoting regional circular economy: a stakeholder perspective. Circul Econ Sustain 5(2):1245–1267. https://doi.org/10.1007/S43615-024-00448-5

Oberoi IS, Rajkumar P, Das S (2021) Disposal and recycling of plastics. Mater Today: Proceed 46:7875–7880. https://doi.org/10.1016/J.MATPR.2021.02.562

OECD (2021) Towards a more resource-efficient and circular economy. OECD Publishing, pp 1–53

OECD (2023) New Aspects of EPR: Extending producer responsibility to additional product groups and challenges throughout the product lifecycle. In: OECD Environment Working Paper No.225. OECD Environment Working Papers, (No. 225), 1–72. www.oecdilibrary.org

OECD (2024) Best available techniques (bat) for preventing and controlling industrial Activity 7 : Cross Country analysis of BAT and BAT-associated emission and and Steel , Paper and Pulp , and Waste. https://www.oecd.org/en/publications/best-available-techniques-bat-for-preventing-and-controlling-industrial-pollution-activity-7-cross-country-analysis-of-selected-brefs-for-iron-steel-paper-pulp-and-waste-incineration-sectors_eff85baf-en.html

OECD (2025) The circular economy in cities and regions of the European Union. https://www.oecd.org/en/publications/the-circular-economy-in-cities-and-regions-of-the-european-union_e09c21e2-en.html

Ogunmakinde OE (2019) Una revisión de los modelos de desarrollo de la economía circular en China, Alemania y Ecuador. Recycling 4(3)

Omer MAE, Mahmoud Ibrahim AM, Elsheikh AH, Hegab H (2025) A framework for integrating sustainable production practices along the product life cycle. Environ Sustain Indicat 26:100606. https://doi.org/10.1016/J.INDIC.2025.100606

Onat NC, Kucukvar M, Halog A, Cloutier S (2017) Systems thinking for life cycle sustainability assessment: a review of recent developments, applications, and future perspectives. Sustainability (Switzerland) 9(5). https://doi.org/10.3390/SU9050706

Paché G (2024) Kalundborg industrial symbiosis: circular strategy in the light of mutualism. J Strateg Innov Sustain 19(3). https://doi.org/10.33423/jsis.v19i3.7330

Patil RA, Langen SKV (2023) Circularity at nano level: a product/service perspective. In: Circularity assessment: macro to nano. Springer Nature Singapore. https://doi.org/10.1007/978-981-19-9700-6

Patil RA, Langen SK, Ramakrishna S (2023) Circularity at micro level: a business perspective. In: Circularity assessment: macro to nano. Springer Natura. https://doi.org/10.1007/978-981-19-9700-6_5

Pinzón-Castro SY, Maldonado-Guzmán G, Rodríguez-González RM (2025) From linear to circular: role of business strategy for green transition in mexican automotive industry. Am J Ind Bus Manag 15(07):949–973. https://doi.org/10.4236/AJIBM.2025.157045

Psarommatis F, Azamfirei V (2024) Zero defect manufacturing: a complete guide for advanced and sustainable quality management. J Manuf Syst 77:764–779. https://doi.org/10.1016/J.JMSY.2024.10.022

Psarommatis F, May G, Azamfirei V (2025) Product reuse and repurpose in circular manufacturing: a critical review of key challenges, shortcomings and future directions. J Remanufact. Springer Netherlands. https://doi.org/10.1007/s13243-025-00153-y

Reichel J (2023) Pollution prevention. Encyclopedia Sustain Manag 2618–2621. https://doi.org/10.1007/978-3-031-25984-5_318

Repair Café (2022) Repara tus objetos rotos. https://www.repaircafe.org/en/. Accessed 6 August 2025

Richards DJ, Fullerton AB (1994) Industrial ecology: US-Japan perspectives. Report on the U.S-Japon Workshop on Industrial Ecology. https://nap.nationalacademies.org/read/9287/chapter/1. Accessed 8 August 2025

Rizos V, Behrens A, van der Gaast W, Hofman E, Ioannou A, Kafyeke T et al (2016) Implementation of circular economy business models by small and medium-sized enterprises (SMEs): barriers and enablers. Sustainability (Switzerland) 8(11). https://doi.org/10.3390/su8111212

Rodríguez-Espíndola O, Cuevas-Romo A, Chowdhury S, Díaz-Acevedo N, Albores P, Despoudi S et al (2022) The role of circular economy principles and sustainable-oriented innovation to enhance social, economic and environmental performance: evidence from Mexican SMEs. Int J Prod Econ 248:108495. https://doi.org/10.1016/J.IJPE.2022.108495

Saari UA, Damberg S, Schneider M, Aarikka-Stenroos L, Herstatt C, Lanz M, Ringle CM (2024) Capabilities for circular economy innovation: factors leading to product/service innovations in the construction and manufacturing industries. J Clean Product 434:140295. https://doi.org/10.1016/j.jclepro.2023.140295

Saavedra YMB, Iritani DR, Pavan ALR, Ometto AR (2018) Theoretical contribution of industrial ecology to circular economy. J Clean Product Elsevier Ltd. https://doi.org/10.1016/j.jclepro.2017.09.260

Sala S, Goralczyk M (2013) Chemical footprint: a methodological framework for bridging life cycle assessment and planetary boundaries for chemical pollution. Integr Environ Assess Manag 9(4):623–632. https://doi.org/10.1002/IEAM.1471

Sekti IS, Singgih ML (2022) Circular economy in plastic pallet manufacturer (PPM) using nano level material circularity indicator (MCI). In: ACM international conference proceeding series: association for computing machinery. https://doi.org/10.1145/3557738.3557881

Shahzabeen A, Ghosh A, Pandey B, Shekhar S (2023) Circular economy and sustainable production and consumption, 43–65. https://doi.org/10.1007/978-3-031-40304-0_3

Da Silveira APP, Mata-Lima H (2020) Energy audit in water supply systems: a proposal of integrated approach towards energy efficiency. Water Policy 22(6):1126–1141. https://doi.org/10.2166/wp.2020.203

Sun Z, Sun X, Wang W, Wang W (2023) Source reduction strategy or end-of-pipe solution? the impact of green merger and acquisition on environmental investment strategy of Chinese heavily polluting enterprises. J Clean Product 413. https://doi.org/10.1016/j.jclepro.2023.137530

Teixeira N (2025) Circular economy perspectives: challenges, innovations, and sustainable futures. Discover Sustain 6(1). https://doi.org/10.1007/S43621-025-01606-X

Tukker A (2015) Product services for a resource-efficient and circular economy–a review. J Clean Prod 97:76–91. https://doi.org/10.1016/J.JCLEPRO.2013.11.049

UNECE (2024) Sustainable development in the UNECE Region: facing a Headwind in 2024. https://unece.org/statistics/publications/sdg2024

UNEP (2006) Environmental agree ments and cleaner production. https://wedocs.unep.org/20.500.11822/7840

UNEP (2012) Sustainable consumption and production: a handbook for policy makers, with cases from Asia and the Pacific. https://www.switch-asia.eu/resource/sustainable-consumption-and-production-a-handbook-for-policy-makers/

UNEP (2024a) Global Waste Management Outlook 2024. (UNEP) United Nations Environment Programme Association, (ISWA) International Solid Waste

UNEP (2024b) Extended producer responsibility. https://www.unep.org/ietc/what-we-do/extended-producer-responsibility. Accessed 8 August 2025

UNIDO (2024) International yearbook of industrial statistics. https://drive.google.com/file/d/1KMBCB7jPlRYqS5WTCtxVwSwFSR0B4pjo/view. Accessed 23 August 2025

Urbinati A, Franzò S, Chiaroni D (2021) Enablers and barriers for circular business models: an empirical analysis in the Italian automotive industry. Sustain Product Consumpt 27:551–566. https://doi.org/10.1016/J.SPC.2021.01.022

Van Berkel R (2011) Evaluation of the global implementation of the UNIDO-UNEP National Cleaner Production Centres (NCPC) Programme. Clean Technol Environ Policy 13(1):161–175. https://doi.org/10.1007/S10098-010-0276-6/METRICS

Velenturf APM, Purnell P (2021) Principles for a sustainable circular economy. Sustain Product Consumpt 27:1437–1457. https://doi.org/10.1016/J.SPC.2021.02.018

Wang G, Feng X, Khim Hoong C (2014) Symbiosis analysis on industrial ecological system. Chin J Chem Eng 22(6):690–698. https://doi.org/10.1016/S1004-9541(14)60084-7

Wang L, Xu X, Gao R, Nee AYC (2019) Sustainable cybernetic manufacturing. Int J Prod Res 57(12):3799–3801. https://doi.org/10.1080/00207543.2019.1598153

Winans K, Kendall A, Deng H (2017) The history and current applications of the circular economy concept. Renew Sustain Energy Rev 68:825–833. https://doi.org/10.1016/J.RSER.2016.09.123

World Bank (2018) What a Waste 2.0: A Global Snapshot of Solid Waste Management to 2050. What a Waste 2.0: A Global Snapshot of Solid Waste Management to 2050. World Bank, Washington, DC. https://doi.org/10.1596/978-1-4648-1329-0

World Bank (2022) Squaring the circle: policies from europe's circular economy transition. https://www.worldbank.org/en/region/eca/publication/squaring-circle-europe-circular-economy-transition

World Economic Forum (2025) Why it's time to transform markets and embrace the circular economy. Circular Economy. https://www.weforum.org/stories/2025/07/age-of-disruption-market-transformation-circular-economy/. Accessed 4 August 2025

Wrage M (2021) Applying lessons learned from water audits. J Am Water Works Assoc 113(3):77–80. https://doi.org/10.1002/awwa.1694

Yuan Z, Bi J (2006) Progress and perspectives of industrial ecology: a critical review. Acta Ecol Sin 26(8):2709–2715

Zhang C, Hu M Di Maio F, Sprecher B, Yang X, Tukker A (2022) An overview of the waste hierarchy framework for analyzing the circularity in construction and demolition waste management in Europe. Sci Total Environ 803. https://doi.org/10.1016/J.SCITOTENV.2021.149892

Zhao H, Guo S, Zhao H (2018) Comprehensive benefit evaluation of eco-industrial parks by employing the best-worst method based on circular economy and sustainability. Environ Dev Sustain 20(3):1229–1253. https://doi.org/10.1007/s10668-017-9936-6

Chapter 4
The Challenge of Downshift Consumerism

Abstract This chapter analyzes the role of consumption in the transition to sustainable models, within the framework of Sustainable Development Goal 12 (SDG 12). It emphasizes that improving production systems is not enough: it is necessary to transform industrial and social consumption patterns. The text reviews the concept of sustainable consumption from its definition at the Rio Summit (1992) and the Oslo Symposium (1994) to its integration into global policies. The chapter examines the environmental and social impact of industrial consumption, especially in sectors such as mining, agriculture, and energy, highlighting how extractivism has altered ecosystems, communities, and cultures. It addresses the drivers of change toward the circular economy: legal frameworks such as the Pollution Prevention Act (PPA); extended producer responsibility policies; technological advances in recycling and energy efficiency; and organizational factors such as leadership, corporate culture, and governance. The chapter also analyzes the role of the consumer, whose environmental awareness and cultural behavior influence the adoption of circular practices. The impulsive consumption model is criticized, and lifecycle thinking is proposed as a tool for responsible decision-making. Finally, it is recognized that industrial transformation is being shaped by the expectations of final consumers, such as millennials and Gen Z, who demand greater transparency, legitimacy, and environmental commitment from companies.

Keywords Sustainable consumption · Extractivism · Energy consumption · Material consumption · Lifecycle thinking · Consumer behavior · Extended producer responsibility · Technological innovation

4.1 Introduction

To truly achieve the goals of SDG 12, it's not enough to make production systems more sustainable—as seen in the previous chapter. It's equally essential to transform the way industry and society consume (Arora and Mishra 2023). According to Raman et al. (2024) moving toward sustainable consumption is strategic for achieving the

L. Velazquez, *Circular Economy and Sustainable Consumption in the Context of SDG 12*, Strategies for Sustainability, https://doi.org/10.1007/978-3-032-15598-6_4

goals of the 2030 Agenda, as it is directly linked to SDG 12 and its synergies with other SDGs such as 13 (climate action), 7 (clean energy), 11 (sustainable cities), and 15 (life on land).

The concept of sustainable consumption was officially introduced in the early 1990s and gained momentum during the 1992 Earth Summit in Rio de Janeiro, where Agenda 21 urged nations to adopt policies that encourage changes in consumption habits (United Nations 1992). However, it was the 1994 Oslo Symposium on Sustainable Consumption that provided one of the most widely accepted definitions: "the use of goods and services that respond to basic needs and bring a better quality of life while minimizing the use of natural resources, toxic materials, and emissions of waste and pollutants over the life cycle, so as not to jeopardize the needs of future generations" (Oslo Symposium 1994). This definition has had a long history of validation by the scientific community. It has been adopted by United Nation Environment Program (UNEP), which has cited it in various documents, including the 10-year framework of programs on sustainable consumption and production (UNEP 2010). In the context of SDG 12, the definition is presented as the conceptual basis for assessing member states' progress toward target 12.1.1 (UNEP 2020).

As highlighted by De et al. (2025), the inclusion of SDG 12 in the global agenda has heightened the urgency of adopting responsible consumption habits. While their review broadly examines consumer behavior and research trends, it underscores the need for interdisciplinary approaches—especially important for large-scale actors whose decisions impact entire supply chains and environmental outcomes. Responsibility for sustainable consumption and production, as mandated by SDG 12, presents a challenge for both industrial and end consumers. However, it is within the industrial context that decision-makers bear the most significant responsibility, as they are embedded in the economic system through their supply chains. Therefore, any decision regarding the material procurement process will have both positive and negative effects that extend beyond the scope of a domestic consumer.

4.2 Industrial Consumption and the Global Acceleration of Material Use

While there have been, since 2011, encouraging signs in industry activity of a decoupling between CO_2 emissions and manufacturing growth, emissions remain at high levels. At the same time, industries still often account for the most significant shares of resource extraction, energy use, and waste production (United Nations Industrial Development Organization [UNIDO] 2024). The following sections examine how these industrial environmental impacts have evolved and why they require urgent attention in the context of global development.

4.2.1 The Resource Extraction

The 2016 Global Material Flows and Resource Productivity report shows that the amount of material mined has tripled, from 22 billion tons in 1970 to nearly 70 billion tons over the 50 years from 1970 to around 2010. A key driver of this increase is industrial activity, which requires substantial quantities of iron, steel, cement, and biomass, depending on the type of activity, such as construction, mining, or manufacturing. The report indicates that per capita extraction rises from 7 to 10 tons in the same period, reflecting improved living standards with better material quality in some areas, but also revealing significant inequalities in access and usage (UNEP International Resource Panel 2016). The report also highlights differences between high-income and low-income regions, which are clearly at odds with the core goal of the 2030 Agenda for Sustainable Development: Leaving no one behind.

From the middle of the twentieth century, and even more since what is called the Great Acceleration, human activity in extractivism has directly affected Earth systems on a global scale, with the International Commission for Stratigraphy suggesting that our current state should be described as a new geological epoch—the Anthropocene (Foster 2024). One point is that, beyond being a technical classification of human-driven planetary change, the Anthropocene represents a shared history: the moment when Homo sapiens—or at least some part of them—became a force capable of significantly disturbing the Earth, to the extent that it transformed one geological epoch into another.

This planetary-scale transformation is not only visible in geological records, but it is also deeply felt in the ecosystems closest to extractive activities. Coal mining impacts far more than the land it affects—it disrupts the equilibrium of the ecosystems surrounding it, with vegetation particularly hard hit, even far away from the mining location. Research has shown that plants surrounding mining areas experience stress and lower levels of vitality due to mining activities (Kayet et al. 2022). Another study found that mines disrupt natural cycles of vegetation over distances far broader than previously known, extending as far as 12 km upslope from the walls of open pits (Liu et al. 2022).

The consequences of extractivism go beyond ecological damage; they also affect the social and cultural fabric of affected communities. Societal and cultural disruption is another effect of mining, as Jacka (2018) demonstrates. She explains that, since the onset of globalization, the demand for new opportunities and resources driven by global capitalism has led to widespread mining expansion, resulting in land dispossession, environmental harm, and significant impacts on the livelihoods of millions, particularly in the Global South. Whether on mountainsides or along riverbeds, these extractive activities—forces that connect people to their land and history—make mining more than just an industry; they turn it into a geophysical force that reshapes the Earth and influences social change. A study by Hanaček et al. (2024) further underscored this problem, explaining how extractivism transcends changing landscapes and severs the long-standing relationships between people and nature. More than just economic or ecological losses, they are losses of identity, memories, and place. Looking across more than 1800 socioecological conflicts, this study demonstrates that the people most affected by extractivism are Indigenous and rural communities whose cultural and spiritual lifeworlds are often torn apart. Consequently, many people have rallied in support not just to safeguard their territory but also to preserve the stories, rituals, and histories that it embodies, and to re-establish dignity through them or to reclaim environmental justice. This growing awareness of extractivism's social and cultural effects has also sparked efforts to reconsider how its impacts are measured and addressed. In response to these concerns, some scholars suggest sustainability frameworks, such as the Mining Area Sustainability Index (MASI), which go beyond economic and environmental stability to enhance the overall well-being of mining communities (Yu et al. 2024).

According to Tang and Werner (2023), the problem is exacerbated by the extent to which the world's mining activity remains hidden. Global, satellite-based assessments have revealed that less than 45% of the world's mining sites are recorded in public production records. However, because data on mining projects is so poorly documented, the lack of transparency is more than just a bureaucratic issue—it also causes environmental damage, illegal activities, and site abandonment without accountability from operators. Small-scale and artisanal mining, which is often informal or unlawful in many regions, largely goes unnoticed. Even legal operations have gaps in data, which can be exploited through corruption or weak governance, making it hard to understand where or how mining occurs.

The ultimate "extractor," mining, appears strongly connected to the cause of the Anthropocene. However, this oversimplifies a complex causal chain. Mining is the final step in a series of industrial and societal needs. It responds to pressures from sectors such as construction, energy, and technology that heavily depend on primary materials. Blaming mining alone overlooks the broader system of consumption and production underlying resource extraction, suggesting that it is not at the core of the problem but rather a reflection of deeper structural forces. A policy brief for the United Nations, "Transforming Extractive Industries for Sustainable Development," outlines the processes by which resources are extracted and processed in the Anthropocene. The report goes beyond mining to detail a range of sectors driven by extractivist logics (United Nations 2021). These include coal, oil, and natural gas in fossil fuel industries; copper, gold, bauxite, and rare earth elements in mineral extraction, and sand, gravel, and clay used in construction materials. Furthermore, the report highlights the growing importance of critical raw materials, such as lithium, cobalt, and graphite, which are essential for our global energy transition. All these share a similar trend: resource-intensive production with significant environmental, social, and governance impacts, mainly affecting countries in the Global South.

Although agriculture is not included in the definition of extractive industries found in the United Nations' Transforming Extractive Industries for Sustainable Development report, other sectors such as mining, fossil fuels, and construction materials are (United Nations 2021). However, in the case of industrial agriculture—particularly large-scale, export-oriented monoculture—this process has come under increasing scrutiny from scholars and grassroots movements that recognize extractivist logics at work. It is precisely these types of platforms that drive the extraction of value from land, labor, and local ecosystems, while ignoring the impact on sustainable livelihoods. These models exhibit the same issues of environmental degradation and social disruption typically seen in the extractive industry itself. The Food and Agriculture Organization of the United Nations (FAO), in its report "State of the World's Land and Water Resources for Food and Agriculture: Options for Sustainable Land Management" (FAO 2021), offers a bleak outlook on the health of global agricultural systems. It presents evidence that the natural resource systems vital for food production are under increasing stress due to overuse, land degradation, and rising competition among different sectors for increasingly limited land and water resources. It also highlights that converting land from agriculture to expand food production leads to significant soil erosion. At the same time, the report concludes that these processes result in declining water quality and ecosystem services. Though it does not explicitly mention monocultures or chemical inputs, it stresses unsustainable farming practices and over-irrigation as two significant causes of resource depletion.

Mining and agriculture, though often seen as separate fields, are increasingly driven by similar extractivist logics in the Anthropocene. Agro-extractivism refers to a form of industrial agriculture that prioritizes large-scale, resource-intensive, and export-oriented production, often achieved through monocultures and the extensive use of chemicals. Like mineral mining, this approach extracts value from land, water, and labor, usually leaving behind degraded ecosystems and marginalized communities. In Kashmir, for example, the expansion of apple monoculture—promoted as a path to prosperity—has increased land use and resource extraction but has not improved smallholder incomes; instead, it has pushed many farmers to the margins (Mushtaq and Goswami 2024). According to Gárgano (2023), in Argentina, agro-extractivist practices have led to environmental health crises as well as socioecological breakdown and reduced rural autonomy. She indicates that higher rates of cancer and respiratory illnesses caused by the expansion of genetically modified soy and increased agrochemical use have fueled grassroots resistance movements that call for agroecological transitions and justice for affected communities. These stories are not isolated—they reflect a global shift in agriculture toward commodification and external control, reinforcing the same patterns of ecological harm and social disruption seen in the mining industry.

4.2.2 *Energy Consumption*

Global energy demand has increased across all regions, but the growth has been uneven, influenced by differences in economic development, climate, and energy policies. According to the Global Material Flows and Resource Productivity Report (UNEP International Resources Panel 2016), worldwide energy demand more than doubled from 1970 to 2010, nearly tripling the amount of material extracted, which increased from 22 billion tons to 70 billion tons. In 2010, fossil fuels remained the primary energy source, making up approximately 80% of the world's energy supply. According to the Energy Institute (2025), North America and Europe experienced the slowest growth rates, at just 0.4% and 0.7%, respectively. Africa saw the smallest absolute increase in energy demand, with only 0.2 exajoules (EJ)—less than half of Europe's increase (0.5 EJ). In contrast, the Asia Pacific region accounted for 68% of the global rise in energy demand and made up 47% of total global consumption. Demand for renewable energy grew by 7%, with China responsible for 56% of that growth—more than the rest of the world combined. Electricity demand continues to grow faster than overall energy demand, with Asia Pacific and the Middle East leading in electricity generation growth at 5.4% and 5.3%, respectively. Both reports converge on informing that the industrial sector has played a central role in driving the surge in global energy demand. This is not merely a technical observation since it reflects data reported by the International Energy Agency (2023), which states that in 2022, the industrial sector consumed approximately 167 exajoules (EJ) out of a total of 442 EJ in global final energy consumption, which corresponds to nearly 38%.

In industries characterized by high energy intensity, such as those in manufacturing, iron and steel, chemicals, non-metallic minerals, non-ferrous metals, and the paper sector, optimizing energy intensity and energy efficiency is crucial for profitability, reducing environmental impact, and improving operational performance. Energy intensity refers to the amount of energy consumed to produce one unit of economic output, typically measured in kilowatt-hours per euro (kWh/EUR) or megajoules per United States dollar (MJ/USD) of gross value added. Energy efficiency refers to the capacity to produce the same output using less energy. In other words, minimize energy consumption without compromising quality or productivity. Inefficiency in energy use is one of the biggest financial wastes in the industry, as there is a clear correlation between low energy efficiency and high operating costs. When a system or facility leaks energy, it requires more fuel or electricity to perform the same work, resulting in a higher energy bill. This additional use not only increases operational costs, but it can also cause extra wear and tear on machines and equipment, which may require more frequent maintenance and a shorter equipment lifecycle (Tsui 2020).

Equally important, reducing energy waste is not only financially wise but also crucial for fighting climate change and lowering humanity's carbon footprint. Addressing both excessive energy consumption and systemic inefficiency in how energy is used is critical to fighting climate change, as both significantly contribute to greenhouse gas emissions. The high demand for energy, particularly from fossil fuels such as coal, oil, and natural gas, directly contributes to increased greenhouse gas emissions, which are a leading cause of human-driven climate change (Wang and Azam 2024). Additionally, populations living in areas with high levels of air pollution are closely linked to significant public health risks, including respiratory and cardiovascular diseases (EPA 2025).

4.2.3 Waste Generation

The level of industrialization, urbanization, and type of economy in countries or regions considerably determines their waste generation rates. As nations develop, both consumption and product diversity increase, increasing the production of not only municipal waste, but also agricultural, construction, industrial, commercial, and healthcare waste. UNEP International Resources Panel (2016) cautions that industries generate massive volumes of solid and liquid waste as well as gaseous emissions and hazardous byproducts. There are different types of industrial waste, such as manufacturing, agricultural, construction and demolition, hazardous, medical, and electronic wastes produced. Industrial waste is derived in a much larger quantity than that of municipal solid waste, nearly 18 times higher than that. This generation is even higher in high-income countries, averaging around 42 kg per person per day. Agricultural waste is also a large proportion, with more than 4.5 times municipal waste (World Bank Group 2018).

There is no moral paradox on earth more distasteful than that of food waste in the face of hunger around the world. Food waste is generated in different sectors, such as fruit and vegetable processing, seafood, dairy, bakery, meat, and ready-to-eat foods, and it can occur at every phase of the supply chain, starting from harvesting, transport, storage, processing, retail, and many times at the consumer end. In agricultural settings, agro-industrial waste can have serious environmental and health effects if it is underutilized or untreated, and ends up incinerated, dumped, or unsafely landfilled, a practice that significantly contributes to rising levels of greenhouse gases such as CO_2 and methane. Furthermore, organic waste generated by the food industry contains large amounts of biological and chemical oxygen demand, as well as suspended solids, which can be released into water bodies, causing health problems for people and animals (Cano-Gómez et al. 2025). The packaging phase is environmentally significant due to its potential to generate waste and emissions. Commercialized food has to be packed for many reasons, including protection during distribution and storage in order to avoid spoilage, and informing the consumer about nutritional, consumption, and health information. The choice of packaging has consequences on the disposal stage, particularly for food and beverage containers, whose low recycling efficiency and high reliance on fossil-based plastics cause environmental pollution and greenhouse gas emission (Operato et al. 2025). Packaging-related environmental issues arise when the design is not conducive to reuse or recycling, for instance, when it is oversized or hard to empty (Frommeyer et al. 2024). Even worse, if food packaging is not well handled, it frequently ends up in waste dumps or on streets and open sites in areas without proper waste management facilities.

The previous information underscores the importance of monitoring and reducing industrial waste to support sustainable industrial growth. However, it is essential to note that waste management also includes waste generated in homes and businesses, which are not treated with the same regulations as industrial waste, but are often just as critical within a circular economy framework. In 2020, the global cost of municipal waste management was $252.3 billion, and it is estimated that, if changes are not implemented, this cost will increase by $165 billion by 2050 due to the continued growth in waste generation (UNEP 2024). Handling a comprehensive waste system is challenging. Literature shows that governments face many constraints in implementing effective waste management policies. Within Organization for Economic Co-operation and Development (OECD), those member countries pursuing a circular economy model are hindered by a lack of economic resources to cover good waste management practices. Some governments have encountered structural and financial difficulties in providing more comprehensive waste management mechanisms. On the other side of the spectrum, Japan has proactively supported the industry and local authorities in developing circular projects and has emerged as one of the leaders of this transition (OECD 2019).

4.3 Sustainability Drivers of Industrial Consumption Reduction

The shift toward a more sustainable industrial consumption is not only an environmental imperative; it is also a strategic opportunity to redefine the value of production. It is within the scope of Sustainable Development Goal 12, which encourages responsible patterns of consumption and production that industries have to meet the challenge of cutting down their material and energy footprint without losing their competitiveness. This section explores the transformational dynamics that drive the momentum to a circular economy, from tighter regulations and shifting societal norms to tech innovations.

4.3.1 Regulatory and Legal Frameworks

In the 70s, laws with a control focus that usually targeted only one medium, such as air, water, or land, forced companies to comply with pollution thresholds. Later, in the mid-1980s and early 1990s, regulatory frameworks such as the Comprehensive Environmental Response, Compensation, and Liability Act (CERCLA) in the United States of America established obligations for industrial companies to report on the use and release of hazardous substances (Stahl et al. 2023), which later evolved into the Pollution Prevention Act. The Pollution Prevention Act was one of the first, if not the first, in the world to establish the foundations for a circular economy by managing waste and contaminants at source by modifying processes, equipment, and products, rather than an end-of-pipe waste management approach (Burnett 1998).

In different parts of the world, albeit to varying levels of enforcement, legal frameworks have transformed to respond to the challenges of promoting circular economy practices as part of their sustainable development agendas. These laws not only regulate waste management and recycling but also incentivize innovation in product design, promote extended producer responsibility, and foster business models that reduce waste. In Europe, North America, Africa, Asia, and Latin America, greater legislative efforts are being made to change the direction of sustainable consumption, which is not only a matter of individual choice but, above all, of a regulatory framework that favors it and allows for systemic viability.

A comparative study of various national legislations to promote a sustainable economy (United Nations Development Programme [UNDP] 2022) claims that China's Circular Economy Promotion Law, enacted in 2008, was one of the first in the world to establish an explicit legal framework for the transition to closed-loop industrial cycles. The legislation also regulates Chinese citizens to enhance circular thinking in their everyday lives. Another illustrative case is that of France, which adopted an ambitious legislation in 2020 that combats waste and promotes circularity at all levels of consumption. German legislation, for its part, has focused its efforts on packaging, with a law that distributes environmental responsibility throughout the supply chain. In Japan, the 2000 Basic Law on Establishing a Recycling-Based Society was the cornerstone for a circular society, where recycling was incorporated in the design and labeling of the product. Similarly, South Korea has adopted a law that protects its environment and encourages the country's society to promote resource circulation. In the Philippines and Taiwan, they have introduced their legal system, which does not directly refer to the circular economy but directly induces recycling, reuse, and sustainability practices to downshift consumerism.

According to the UNEP (2015), others have opted for more flexible but equally meaningful paths: national plans, sustainable development strategies, and sectoral policies that reflect a commitment to sustainable consumption and production. For instance, Brazil, Colombia, Ecuador, Ghana, Indonesia, and Zambia have developed dedicated sustainable consumption and production plans, which focus on waste reduction, resource conservation, and responsible public procurement. In Europe, the European Commission launched an action plan in 2008 linking responsible consumption to industrial policy. Canada implemented influential sectoral policies to foster green procurement policy, which affects billions of dollars in public procurement. In the United States of America, the Environmental Protection Agency published a guide for Green Purchasing to assist the public and private interested parties in procuring green products.

Another interesting and increasingly popular approach to encourage the collection and recovery of materials at the post-consumer stage is Extended Producer Responsibility (EPR). Introduced in the early 1990s in Sweden, EPR significantly changed how key actors handle waste management by making manufacturers responsible for the entire lifecycle of their products (OECD 2024). Today, this approach is being gradually adopted—sometimes voluntarily, sometimes through regulation—in various countries around the world (Castrillon-Ocampo et al. 2025; OECD 2024; Portugaise et al. 2023).

In the literature, there is a lot of evidence of worldwide cases that it is almost impossible and impractical to attempt to inventory the international compendium of legal frameworks, laws, and policies linked to encouraging sustainable consumption patterns in industry. However, the national cases included in this section help exemplify how legislation works in different contexts. While diverse in approach, they converge in the common goal of transforming consumption into a more conscious and regenerative practice.

4.3.2 Science and Technology

The development of sustainability science and new environmentally friendly technologies has promoted the move toward circular and efficient systems. Sustainability science is described as a mature, transdisciplinary, applied, challenge-driven field to guide interventions in problems emanating from human–nature interaction (Fang et al. 2018). As Schirone (2025) suggests, the historical development of sustainability science can be summarized in three dominant periods, which illustrate essential dynamics. The first period, which he refers to as the foundational, is marked by the growth of a scientific community with a concern for addressing socio-environmental problems from a transdisciplinary angle; here, systems thinking, participatory research, and indigenous knowledge are emphasized. Then comes the introspective period, when the theoretical bases and the methods are debated, more specifically in what concerns the vehicle of transformation in the productive system. At the most developed period, what is called the diversification, there is an aperture to advance innovative solutions to solve unsustainability problems in diverse fields with the active involvement of non-academic actors such as policymakers, the private sector, and local communities.

Sustainability science challenges the dominant norms and practices of established disciplines of economics, ecology, and engineering. Its heterodox character has opened space for paths of systematic transformation with an emphasis on technology. Technological innovation, therefore, has a central role to play in enhancing resource efficiency and waste reduction, making it a powerful tool in the transition toward the circular economy, in line with Sustainable Development Goal 12. The World Bank, in the report Squaring the Circle: Policies from Europe's Circular Economy Transition, highlights technological innovation as a fundamental driver in the transformation toward a circular economy, as it allows for the redesign of materials, optimization of production and reuse processes, and the application of digital tools that improve resource efficiency (World Bank 2022). Advanced recycling technologies, intelligent product design, and the use of digital data have proven to be key to increasing productivity and reducing waste. It also states that there is evidence of a direct relationship between resource efficiency and productivity gains, driven primarily by the level of technological innovation. Furthermore, it suggests that the European Union can leverage global value chains to disseminate these technologies and achieve material efficiency improvements in production processes outside of Europe, while limiting the risks of resource leakage. Modern waste treatment facilities have been developed along technological innovations that tackle the environmental impacts in lifecycle stages of the waste, which vary from its generation, collection, transportation logistics, and processing. To enable applications of zero-waste technologies in various processes, a strong and unified governance system across governments, businesses, and communities is needed for the effective and sustainable implementation (Kumar et al. 2025). In this sense, the use of digital governance platforms that combine WebGIS and smart sensors has enabled more efficient, transparent, and participatory management in the form of disruptive innovation (Rittl et al. 2025). Recycling methods such as mechanical, thermolysis, chemical depolymerization processes, and biological processes facilitate the valorization of plastic materials post-consumer usage. The integrated artificial intelligence and big data analytics have enhanced the efficiency and optimized the cost operations and material traceability, which have played an appreciable role in promoting a more innovative and sustainable circular economy (Jiang and Bateer 2025). According to Czekała et al. (2023), robotics is an efficient tool for improving waste sorting efficiency. The authors note that robotic systems, especially those equipped with computer vision and machine learning algorithms, make it possible to identify and separate materials more accurately than manual methods, increasing recovery rates and reducing cross contamination in recycling streams. In addition, with the use of blockchain technologies, it is possible to enhance trust and transparency in the waste management lifecycle by providing data integrity and secure traceability along the lifecycle, giving citizens the ability to track the movement of waste from its first collection to new products (Vorobeva et al. 2023).

Whether implemented in conventional waste management systems or through digital platforms that promote transparency and stakeholder collaboration, technology plays a pivotal role in the transition toward circularity. Rather than acting as isolated tools of sustainability science, technological innovations function as an integrated component within a transdisciplinary framework. This framework catalyzes the revalorization of waste by enabling closed-loop systems, optimizing resource recovery, and fostering adaptive governance models that align with zero-waste objectives.

4.3.3 Organizational Factors

In addition to the legal framework and scientific and technological advances, organizational factors are essential for moving toward more sustainable industrial consumption. From a strategic perspective, corporate sustainability refers to the integration of elements from the three dimensions of sustainability—environmental, social, and economic—into the ultimate corporate purpose (Florez-Jimenez et al. 2025). It comprises decision-making at the management level that directly impacts the entire organizational structure, affecting resource use and environmental pollution. It thus determines sustainability performance, which refers to how well these principles are put into practice—via tangible results such as resource use, emissions, and social development. In such a sense, sustainability performance represents a concrete manifestation of a firm's dedication to corporate sustainability.

Corporate sustainability is significantly affected by organizational culture, leadership, corporate social responsibility, transparency, and other internal elements. These organizational factors shape the values and behaviors of employees, managers, and even suppliers. In this sense, effective business leadership is needed to promote and integrate individual and collective actions that align with the vision, mission, and long-term sustainability goals of the corporation (Lizares and Cuyegkeng 2024). While this is generally the case, there are instances where the influence may be bidirectional; that is, individual values and behaviors can also shape organizational culture. Therefore, effective business leadership should integrate individual and collective actions in a way that reinforces corporate sustainability.

This system of norms, policies, procedures, and practices that govern a company is essential in creating a foundation for aligning the organization's policies, norms, and procedures with its broader sustainability principles (Hristov et al. 2022). Significantly, to effectively strengthen the internal conditions needed to embed sustainability values into the organizational culture, governance must promote transparency and accountability (Alemu 2025). With the increasing importance of disclosure, companies are more frequently relying on sustainable reporting frameworks to track their progress and hold themselves accountable. The Global Reporting Initiative (GRI), International Organization for Standardization (ISO) 14001, and the Eco-Management and Audit Scheme (EMAS) are established and proven schemes that provide guidelines for measuring and reporting sustainability (de Junguitu and Allur 2019; Rivo-López et al. 2025). Recently, there has been a growing emphasis on developing skilled human talent capable of integrating sustainability indicators into corporate governance by Environmental, Social, and Governance (ESG) reporting, which is a novel framework that reflects a shift in how companies assess corporate sustainability (Cai et al. 2024). The ESG framework incorporates social indicators such as labor treatment, equity, inclusion, and community outreach in order to establish stronger relational links with corporations, stakeholders, and local community members while enhancing corporate reputation and building trust (Omonijo and Zhang 2025).

While sustainability reporting frameworks have traditionally been voluntary tools, their role is evolving due to increasing regulatory pressure. This link between internal governance and corporate sustainability has gained legal support through Directive (EU) 2024/1760, on corporate sustainability due diligence, which came into effect on July 25, 2024. It establishes a mandatory legal framework for large companies to identify, prevent, and mitigate negative impacts on human rights and the environment in their operations and supply chains, both within and outside Europe. The directive sets mandatory requirements for companies to recognize, assess, prevent, reduce, and address adverse impacts on human rights and the environment, aligning directly with the social and environmental components of the ESG framework. It mandates, among other things, the adoption of a climate transition plan aligned with the 2050 climate neutrality goal, as well as the implementation of monitoring, transparency, and redress mechanisms. The directive applies to EU companies with more than 1,000 employees and a turnover exceeding €450 million, as well as to non-European companies with a significant presence in the European Union market (European Union 2024).

Organizational factors are only one of the main drivers of change; their structural and operational effects can significantly facilitate or hinder the implementation of new strategies aimed at reducing industrial consumption. Given this, it is also essential to consider cultural factors, as their symbolic and social dimensions add complexity to analyzing market pressures that influence consumption by evolving consumer expectations, societal values, and collective attitudes toward circular economy practices.

4.3.4 Cultural Factors

Another critical aspect of advancing the circular economy lies in cultural factors that shape consumer behavior (Zavos et al. 2024). Final consumers are economic entities that purchase goods or services for their personal or group use. They exert a powerful influence in the market that can drive industry production toward responsible consumption, but this demands a paradigm shift in purchasing behavior, which currently doesn't exist in many communities, mainly where consumers tend to buy compulsively. According to Sofi et al. (2020), impulsive buying behavior is determined by an interaction between cognitive and affective processes, where decisions are made with low rational deliberation and high emotional influence. This consumption pattern, guided by immediate stimuli and instant gratification, leads to a phenomenon called consumerism, which represents a significant obstacle to the adoption of sustainable consumption practices.

Social media platforms further exacerbate consumerism, particularly among younger consumers, by promoting impulsive buying behaviors and idealized lifestyles that reinforce short-term satisfaction (Ciocodeică et al. 2025). Known for their heavy use of social media, millennial and Gen Z consumers tend to be proactive in their pursuit of sustainable products; however, millennials prioritize their ethical values when purchasing eco-friendly products, even if it means paying a higher price—a behavior that Gen Z generally doesn't exhibit. While Gen Z also values sustainability, they often prioritize factors such as fashion, price, or influencer opinions, which can lead them to prefer products that are not as sustainable (Savin et al. 2024).

Consumerism can be defined as the belief that it is good for people to spend a lot of money on goods and services (Britannica Dictionary 2025). In economic terms, consumption refers to the total expenditure by households on final goods and services within a given period (Wahlen and Laamanen 2015). It constitutes a core component of Gross Domestic Product (GDP) under the expenditure approach, reflecting the level of domestic demand and serving as a key indicator of economic activity and welfare (Barroso Campos 2011; Blanchard et al. 2012). Therefore, in financial terms, higher levels of household consumption are generally associated with increased GDP growth, as they stimulate greater production and sales of goods and services within the economy (Bergh 2009). As it is a key driver of gross domestic product growth and production profitability, both governments and industry need to promote consumerism (Bodislav and Georgescu 2025), whether through advertising or public policies (Mont et al. 2022). Although the contribution of consumerism to gross domestic product is clearly measurable through consumer spending, this approach fails to incorporate environmental externalities, which pose significant threats to genuine human well-being (Aitken 2019). Notably, the negative externalities often remain invisible to the average consumer unless they possess prior knowledge or a critical understanding of systemic impacts.

Consumer knowledge and awareness are essential drivers in the acceptance of circular solutions, as they enable individuals to understand the long-term benefits and environmental implications of sustainable alternatives (Camacho-Otero et al. 2018). These environmental externalities include environmental degradation, resource exploitation, energy consumption, and waste generation, which are embedded throughout the value chain but obscured by marketing narratives and consumption habits. As Reyes-García et al. (2025) point out, many environmentally harmful activities are sustained by subsidies that distort market signals and hide their actual ecological costs. This institutional support perpetuates unsustainable practices and makes it harder for consumers to perceive the environmental impact of their choices.

To reveal production's externalities, individuals can critically evaluate their purchasing choices, reflect on their consumption habits, and increase awareness of the environmental impacts associated with the products they buy. Lifecycle thinking supports this by helping consumers understand how their decisions influence each stage of a product's life, encouraging more responsible and sustainable consumption. Lifecycle thinking is a tool recommended for sustainability scholars to inform consumers about the environmental externalities of products (Tsuda et al. 2013). Sustainability-minded consumers, guided by lifecycle thinking, can identify the hidden environmental impacts at each stage of a product's life (Lewandowska et al. 2018). As consumers become more aware of the negative externalities of consumption, they are more engaged in purchasing responsible products that reduce waste and maximize the use of energy, water, and materials (Kutaula et al. 2024).

To advance toward sustainable consumption in the context of SDG 12, a consumer should no longer be seen as a passive recipient of products but as an informed and active participant driving meaningful changes that indirectly facilitate the adoption of circularity practices into industrial production processes (Henriques et al. 2023). A critical aspect for the advancement of the circular economy is consumer behavior, shaped by cultural factors that define their buying preferences (Sahai Meena 2023). Therefore, active consumer participation is essential for the success of circular initiatives such as the return of packaging or cans or the purchase of products made from recycled materials (Richard 2024).

4.4 The Evolving Role of Industry in Sustainable Consumption

Industrial sectors around the world are increasingly acknowledging the need to align their processes and practices with Sustainable Development Goal 12. A comparative efficiency assessment points out that the OECD countries have paved the way in the process of adopting the practice of sustainable consumption and production due to the policy environment, which is policy-framed, integrated, and technological development (Castellano et al. 2024). These transformations are not restricted to any one sector, but signal a systemic change in the way industry views the circular economy's principles. More and more, firms understand that sustainability is not only about legality and reputation, but rather a legitimacy and in many cases a profitability issue. Under this approach, the circular economy and sustainable consumption appear to be the main drivers to steer industries toward new models that break the relationship between economic development, environmental degradation, and social burden. Yet, in times of higher public oversight and ecological sensitivity, though, it may be necessary to examine whether such corporate efforts are genuinely binding or a strategic branding effort, the dilemma of commitment versus greenwashing.

As mentioned previously in this chapter, mining is ultimately about the extraction of non-renewable natural resources, but, on the other hand, it is one of the most proactive industries that seek to minimize its environmental impact and maximize social benefits. According to Pavloudakis et al. (2024), through the analysis of 17 studies, mining companies are conducting significant efforts to minimize their ecological footprint, mainly through the implementation of sustainable mining practices, such as the use of clean technologies, the electrification of heavy machinery, the implementation of renewable energy in operational processes, the optimization of water consumption through closed systems, and the application of circular economy models to reduce waste. Law and enforcement are also key drivers for sustainability in the mining industry. According to the authors, legal frameworks and strict enforcement of environmental laws have encouraged enterprises to invest in clean technologies and deliver better waste management and community participation. Furthermore, circular economy's practices have transformed the concept of mining, evolving it into a strategy for recovering and revalorizing buried materials as useful resources. This approach aims to extract metals, plastics, fuels, and other usable materials from old landfills. However, the results have generally fallen short of economic and technical expectations (Krook et al. 2012).

According to Scharmer et al. (2024), sustainable manufacturing can be viewed as a strategic concept based on the triple bottom line. From an environmental standpoint, the range of sustainability practices is extensive, including indicators such as efficient use of energy, water, and natural resources. At the same time, they allocate funding to promote recycling, reduce toxic substances, and prevent and control emissions that could impact air quality or other areas. From a social perspective, it promotes corporate social responsibility, decent working conditions, and community involvement. Finally, the economic aspect aims for the financial sustainability of these practices by encouraging technological innovation and regulation, which in turn fosters competition. This holistic approach helps sectors transition toward sustainable and responsible production systems that align with global sustainability goals. For their part, Setyadi et al. (2025) support the argument that the concept of sustainable manufacturing has evolved around the Triple Bottom Line approach, highlighting a strong trend toward the circular economy and practices focused on reducing energy consumption and material waste. In economic terms, the adoption of emerging technologies such as artificial intelligence, blockchain, and digital twins stands out, which have been shown to reduce energy costs and improve operational efficiency. However, the authors emphasize that the social dimension remains the least addressed, with gaps in inclusive policies, labor equity, and community engagement.

Years ago, industry started an ongoing and profound transformation by adopting the circular economy as an essential part of its journey toward sustainability. Managers and supervisors are reengineering processes and operations to no longer produce more, but to do so in a sustainable manner. The adoption of digital technologies such as artificial intelligence and digital twins is helping to create more efficiently and with less environmental impact. Interestingly, this evolution is not only a response to regulations or external pressures, but also to the desire of many companies to genuinely be part of the solution (Tukker and Dietzenbacher 2013). Finally, it is worth noting that this ongoing transformation in the industry does not occur in isolation; it is actively shaped by the values and expectations of young household consumers, especially Generation Z, who are more aware of environmental externalities and are helping to define what responsible consumption truly means (Liang et al. 2024).

4.5 Conclusion

Transforming consumption is a central challenge in the pursuit of sustainability. This chapter has shown that industrial and societal consumption patterns are deeply intertwined with extractivist logics, environmental degradation, and social disruption. Addressing these issues requires more than technical adjustments—it demands a cultural and structural shift in how value is created, distributed, and perceived.

Legal frameworks, technological innovation, and organizational governance have emerged as key drivers in reducing industrial consumption and promoting circularity. However, the role of consumers is equally critical. Their awareness, behavior, and expectations shape market dynamics and influence industrial transformation. Moving beyond impulsive consumerism toward informed, responsible choices is essential for aligning production systems with ecological limits and social equity.

Ultimately, sustainable consumption is not only about reducing material and energy footprints—it is about redefining prosperity, restoring dignity to communities affected by extractivism, and fostering a culture of care and accountability. The transition to circular models must be guided by ethical commitment, interdisciplinary collaboration, and inclusive participation across all sectors of society.

References

Aitken A (2019) Measuring welfare beyond GDP. Natl Inst Econ Rev 249(1):R3–R16. https://doi.org/10.1177/002795011924900110

Alemu BA (2025) Cultivating a culture of sustainability: the role of organizational values and leadership in driving sustainable practices. Bus Ethics Leadership 9(1):79–94. https://doi.org/10.61093/bel.9(1).79-94.2025

Arora NK, Mishra I (2023) Responsible consumption and production: a roadmap to sustainable development. Environ Sustain 6(1):1–6. https://doi.org/10.1007/S42398-023-00266-9

Barroso Campos C (2011) Fundamentos de Economía. Revista de Fomento Social 583–589. https://doi.org/10.32418/rfs.2011.263.1858

Blanchard O, Amighini A, Giavazzi F (2012) Macroeconomía (5ta ed.). Pearson Educación, S.A, Madrid

Bodislav DA, Georgescu RI (2025) Consumerism and capitalism: dynamic evolution, economic benefits, and ethical challenges. Theor Appl Econ XXXII(1):251–266

Britannica Dictionary (2025) Definition of consumerism. Encyclopædia Britannica, Inc. https://www.britannica.com/dictionary/consumerism. Accessed 5 Sept 2025

Burnett ML (1998) The pollution prevention act of 1990: a policy whose time has come or symbolic legislation? Environ Manage 22(2):213–224. https://doi.org/10.1007/s002679900098

Cai X, Xiang H, Neskorodieva I, Durmanov A (2024) Interrelation between human capital management and ESG engagement: evidence from S&P 500 firms. Humanities Soc Sci Commun 11(1). https://doi.org/10.1057/S41599-024-04189-6

Camacho-Otero J, Boks C, Pettersen IN (2018) Consumption in the circular economy: a literature review. Sustainability (Switzerland) 10(8). https://doi.org/10.3390/SU10082758

Cano-Gómez CI, Wong-Arguelles C, Hinojosa-López JI, Muñiz-Márquez DB, Wong-Paz JE (2025) Novel Insights into agro-industrial waste: exploring techno-economic viability as an alternative source of water recovery. Waste 3(2):15. https://doi.org/10.3390/WASTE3020015

Castellano R, De Bernardo G, Punzo G (2024) Sustainable well-being and sustainable consumption and production: an efficiency analysis of sustainable development goal 12. Sustainability (Switzerland) 16(17). https://doi.org/10.3390/SU16177535

Castrillon-Ocampo J, Jaén S, Maya-Duque P (2025) A system dynamics prospective analysis of extended producer responsibility policies: a business for a few. Circ Econ Sustain 3317–3342.https://doi.org/10.1007/s43615-025-00547-x

Ciocodeică DF, Chivu RG, Popa IC, Mihălcescu H, Barghier I (2025) Hedonic and impulsive consumer behavior stimulated by social media: implications for sustainable fashion marketing. Sustainability (Switzerland) 17(11). https://doi.org/10.3390/SU17115198

Czekała W, Drozdowski J, Łabiak P (2023) Modern technologies for waste management: a review. Appl Sci (Switzerland) 13(15). https://doi.org/10.3390/app13158847

De T, Singh N, Ibrahim BSKSMK, Dikkatwar R, Chaudhary MK (2025) Mindful, responsible, and sustainable consumption: a hybrid review of evolving research trends. Discover Sustain 6(1). https://doi.org/10.1007/s43621-025-01343-1

de Junguitu AD, Allur E (2019) The adoption of environmental management systems based on ISO 14001, EMAS, and alternative models for SMEs: a qualitative empirical study. Sustainability (Switzerland) 11(24). https://doi.org/10.3390/SU11247015

Energy Institute (2025) Statistical review of world energy.pdf—google drive. https://drive.google.com/file/d/1bb4dQEHukLo_rSqyKiho3bnBkP-Ad25H/view. Accessed 23 Aug 2025

EPA (2025) Research on health effects from air pollution. Air Res. https://www.epa.gov/air-research/research-health-effects-air-pollution. Accessed 23 Aug 2025

European Union (2024) Directive (EU) 2024/1760, of 13th June 2024: on corporate sustainability due diligence and amending directive (EU) 2019/1937 and regulation (EU) 2023/2859. Off J Eur Union 1760(3):1–58. http://data.europa.eu/eli/dir/2024/1619/oj

Fang X, Zhou B, Tu X, Ma Q, Wu J (2018) "What kind of a science is sustainability science?" An evidence-based reexamination. Sustainability (Switzerland) 10(5). https://doi.org/10.3390/su10051478

FAO (2021) The state of the World's land and water resources for food and agriculture—systems at breaking point. Rome. https://doi.org/10.4060/cb7654en

Florez-Jimenez MP, Lleo A, Ruiz-Palomino P, Muñoz-Villamizar AF (2025) Corporate sustainability, organizational resilience, and corporate purpose: a review of the academic traditions connecting them. RMS 19(1):67–104. https://doi.org/10.1007/S11846-024-00735-3

Foster JB (2024) Extractivism in the anthropocene. Monthly Rev 1–22. https://doi.org/10.14452/MR-073-11-2022-04_1

Frommeyer B, Koch J, Scagnetti C, Lorenz M, Schewe G (2024) Recycled or reusable: a multimethod assessment of eco-friendly packaging in online retail. J Ind Ecol 28(1):100–115. https://doi.org/10.1111/JIEC.13447

Gárgano C (2023) Agroextractivism in Argentina environmental health, scientific agendas, and socioecological crisis. Front Public Health 11. https://doi.org/10.3389/fpubh.2023.1304514

Hanaček K, Tran D, Landau A, Sanz T, Thiri MA, Navas G et al (2024) "We are protectors, not protestors": global impacts of extractivism on human–nature bonds. Sustain Sci 19(6):1789–1808. https://doi.org/10.1007/S11625-024-01526-1

Henriques R, Figueiredo F, Nunes J (2023) Consumers' perspectives on circular economy: main tendencies for market valorization. Sustainability (Switzerland) 15(19). https://doi.org/10.3390/SU151914292

Hristov I, Chirico A, Ranalli F (2022) Corporate strategies oriented towards sustainable governance: advantages, managerial practices and main challenges. J Manage Governance 26(1):75–97. https://doi.org/10.1007/s10997-021-09581-x

International Energy Agency (2023) World energy outlook 2023 | enhanced reader. World Energy Outlook 2023:23–28. https://www.iea.org/news/the-energy-world-is-set-to-change-significantly-by-2030-based-on-today-s-policy-settings-alone

Jacka JK (2018) Annual review of anthropology: the anthropology of mining: the social and environmental impacts of resource extraction in the mineral age. Annu Rev Anthropol 47:61–77. https://doi.org/10.1146/annurev-anthro-102317-050156

Jiang X, Bateer B (2025) A systematic review of plastic recycling: technology, environmental impact and economic evaluation. Waste Manage Res.https://doi.org/10.1177/0734242X241310658/ASSET/83ABDC61-14BD-4A07-85D7-8DC0BFC4BC9F/ASSETS/IMAGES/LARGE/10.1177_0734242X241310658-FIG7.JPG

Kayet N, Pathak K, Singh CP, Chowdary VM, Bhattacharya BK, Kumar D et al (2022) Vegetation health conditions assessment and mapping using AVIRIS-NG hyperspectral and field spectroscopy data for environmental impact assessment in coal mining sites. Ecotoxicol Environ Saf 239(November 2021):113650. https://doi.org/10.1016/j.ecoenv.2022.113650

Krook J, Svensson N, Eklund M (2012) Landfill mining: a critical review of two decades of research. Waste Manage 32(3):513–520. https://doi.org/10.1016/j.wasman.2011.10.015

Kumar R, Gupta G, Hussain A, Rani A, Thapliyal A, Gunsola D et al (2025) Pioneering zero-waste technologies utilization and its framework on sustainable management: international, national and state level. Discover Appl Sci 7(3). https://doi.org/10.1007/s42452-025-06693-z

Kutaula S, Gillani A, Gregory-Smith D, Bartikowski B (2024) Ethical consumerism in emerging markets: opportunities and challenges. J Bus Ethics 191(4):651–673. https://doi.org/10.1007/S10551-024-05657-4

Lewandowska A, Witczak J, Giungato P, Dierks C, Kurczewski P, Pawlak-Lemanska K (2018) Inclusion of life cycle thinking in a sustainability-oriented consumer's typology: a proposed methodology and an assessment tool. Sustainability (Switzerland) 10(6). https://doi.org/10.3390/SU10061826

Liang J, Li J, Cao X, Zhang Z (2024) Generational differences in sustainable consumption behavior among Chinese residents: implications based on perceptions of sustainable consumption and lifestyle. Sustainability (Switzerland) 16(10). https://doi.org/10.3390/su16103976

Liu Y, Zhou W, Yan K, Guan Y, Wang J (2022) Identification of the disturbed range of coal mining activities: a new land surface phenology perspective. Ecol Ind 143(August):109375. https://doi.org/10.1016/j.ecolind.2022.109375

Lizares REJR, Cuyegkeng MAC (2024) Building sustainable organizations online: How leaders embed sustainability into organizational culture in virtual contexts. Sustain Dev 32(6):6434–6448. https://doi.org/10.1002/SD.3036

Mont O, Lehner M, Dalhammar C (2022) Sustainable consumption through policy intervention—a review of research themes. Front Sustain 3. https://doi.org/10.3389/FRSUS.2022.921477/FULL

Mushtaq U, Goswami K (2024) Agro-extractivism and shifting smallholder dynamics: evidence from Kashmir, India. Dev Stud Res 11(1):1–16. https://doi.org/10.1080/21665095.2024.2426653

OECD (2019) Waste management and the circular economy in selected OECD countries. Evidence from environmental performance reviews. OECD.https://doi.org/10.1787/9789264309395-EN

OECD (2024) Extended producer responsibility, vol 41. https://doi.org/10.1787/67587B0B-EN

Omonijo ON, Zhang Y (2025) Examining the relationship between technological innovation, environmental social governance and corporate sustainability: the moderating role of green operational innovation. HumIties Soc Sci Commun 12(1). https://doi.org/10.1057/S41599-025-04389-8

Operato L, Panzeri A, Masoero G, Gallo A, Gomes L, Hamd W (2025) Food packaging use and post-consumer plastic waste management: a comprehensive review. Front Food Sci Technol 5https://doi.org/10.3389/FRFST.2025.1520532

Pavloudakis F, Roumpos C, Spanidis PM (2024) Sustainable mining and processing of mineral resources. Sustainability (Switzerland) 16(19). https://doi.org/10.3390/SU16198393

Portugaise MK, Jóhannsdóttir L, Murakami S (2023) Extended producer responsibility's effect on producers' electronic waste management practices in Japan and Canada: drivers, barriers, and potential of the urban mine. Discov Sustain 4(1). https://doi.org/10.1007/s43621-023-00124-y

Raman R, Lathabai HH, Nedungadi P (2024) Sustainable development goal 12 and its synergies with other SDGs: identification of key research contributions and policy insights. Discov Sustain 5(1). https://doi.org/10.1007/S43621-024-00289-0

Reyes-García V, Villasante S, Benessaiah K, Pandit R, Agrawal A, Claudet J et al (2025) The costs of subsidies and externalities of economic activities driving nature decline. Ambio 54(7):1128–1141. https://doi.org/10.1007/S13280-025-02147-3

Richard R (2024) Circular economy innovations : transforming waste into resources 09:3–4. https://doi.org/10.37421/2475-7675.2024.9.355

Rittl LGF, Zaman A, de Oliveira FH (2025) Digital transformation in waste management: disruptive innovation and digital governance for zero-waste cities in the global south as keys to future sustainable development. Sustainability (Switzerland) 17(4):1–20. https://doi.org/10.3390/su17041608

Rivo-López E, Villanueva-Villar M, Pardo-López C, Cortés-García M (2025) Environmental sustainability reporting: a systematic and bibliometric review of two decades of research. Futur Bus J 11(1). https://doi.org/10.1186/S43093-025-00580-0

Sahai Meena B (2023) The effect of cultural factors on consumer behaviour: a global perspective. Int J Multidiscip Res 5(6):1–7. https://doi.org/10.36948/ijfmr.2023.v05i06.10906

Savin P-S, Rusu G, Miu C-M, Ciocodeică D-F, Kasem EIFT (2024) Generational perspectives on sustainable consumption: exploring consumer behaviour of millennials and generation Z. In: 7th International conference on economics and social sciences, pp 380–391. https://doi.org/10.24818/icess/2024/038

Scharmer VM, Vernim S, Horsthofer-Rauch J, Jordan P, Maier M, Paul M et al (2024) Sustainable manufacturing: a review and framework derivation. Sustainability (Switzerland) 16(1). https://doi.org/10.3390/SU16010119

Schirone M (2025) The emergence of sustainability science in the editorials of three scholarly journals. Discov Sustain 6(1). https://doi.org/10.1007/s43621-025-01519-9

Setyadi A, Soekotjo S, Lestari SD, Pawirosumarto S, Damaris A (2025) Trends and opportunities in sustainable manufacturing: a systematic review of key dimensions from 2019 to 2024. Sustainability (Switzerland) 17(2). https://doi.org/10.3390/SU17020789

Sofi SA, Mir FA, Baba MM (2020) Cognition and affect in consumer decision making: conceptualization and validation of added constructs in modified instrument. Futur Bus J 6(1). https://doi.org/10.1186/S43093-020-00036-7

Stahl RG, Martin J, Tomasi T, Goldsmith BJ (2023) If coordination of remediation and restoration under CERCLA is such a good idea, why is it not practiced more widely? J Environ Manage 340(May):117964. https://doi.org/10.1016/j.jenvman.2023.117964

Oslo Symposium (1994) Definición del consumo sostenible. Mesa Redonda de Oslo sobre Producción y Consumo Sostenibles. https://enb.iisd.org/consume/oslo004.html. Accessed 28 Aug 2025

Tang L, Werner TT (2023) Global mining footprint mapped from high-resolution satellite imagery. Commun Earth Environ 4(1):1–12. https://doi.org/10.1038/S43247-023-00805-6;SUBJMETA=172,213,2151,4081,4115,685,704,844;KWRD=ECONOMIC+GEOLOGY,ENVIRONMENTAL+IMPACT,SUSTAINABILITY

Tsuda K, Hara K, Uwasu M (2013) Prospects and challenges for disseminating life cycle thinking towards environmental conscious behaviors in daily lives. Sustainability (Switzerland) 5(1):123–135. https://doi.org/10.3390/SU5010123

Tsui J (2020) Three ways our outdated infrastructure contributes to climate change. https://www.renewableenergymagazine.com/jenna-tsui/three-ways-our-outdated-infrastructure-contributes-to-20200804. Accessed 23 Aug 2025

Tukker A, Dietzenbacher E (2013) Global multiregional input-output frameworks: an introduction and outlook. Econ Syst Res 25(1):1–19. https://doi.org/10.1080/09535314.2012.761179

UNDP (2022) Estudio Comparado de Legislación Internacional sobre Economía Circular—Zona de Conocimiento. https://pnud-conocimiento.cr/repositorio/estudio-economia-circular/. Accessed 23 Aug 2025

UNEP International Resource Panel (2016) Global material flows and resource productivity: assessment report for the UNEP international resource panel. https://www.unep.org/resources/report/global-material-flows-and-resource-productivity-assessment-report-unep. Accessed 23 Aug 2025

UNEP International Resources Panel (2016) Global material flows and resource productivity. https://www.resourcepanel.org/es/informes/Enlace-a-la-base-de-datos-de-flujos-de-materiales-globales-y-productividad-de-recursos

UNEP (2010) ABC of SCP: clarifying concepts on sustainable consumption and production. United Nations Environ Program. http://www.uneptie.org/scp/marrakech/pdf/ABCofSCP-ClarifyingConceptsonSCP.pdf%0Ahttps://sustainabledevelopment.un.org/index.php?page=view&type=400&nr=945&menu=1515

UNEP (2015) Sustainable consumption and production: a handbook for policymakers, global edition. United Nations Environ Program 144(3717):1245

UNEP (2020) Metadata sustainable development goals: SDG indicator 12.1.1. Kementerian PPN/Bappenas (141):1–133

UNEP (2024) Global Resources Outlook 2024: Bend the trend—pathways to a liveable planet as resource use spikes. Int Resour Panel. https://wedocs.unep.org/items/3ec5991f-1ab2-4d54-9000-c9d3cee81d16

UNIDO (2024) International Yearbook of Industrial Statistics. https://drive.google.com/file/d/1KMBCB7jPlRYqS5WTCtxVwSwFSR0B4pjo/view. Accessed 23 Aug 2025

United Nations (1992) Agenda 21. Izvestiya—Akademiya Nauk, Seriya Geograficheskaya 6(June):47–54.https://doi.org/10.4135/9781412971867.n128

United Nations (2021) Transforming extractive industries for sustainable development. (May):1–18. https://www.worldbank.org/en/topic/extractiveindustries/overview.%0Ahttps://www.un.org/sites/un2.un.org/files/sg_policy_brief_extractives.pdf

van den Bergh JCJM (2009) The GDP paradox. J Econ Psychol 30(2):117–135. https://doi.org/10.1016/J.JOEP.2008.12.001

Vorobeva D, Scott IJ, Oliveira T, Neto M (2023) Leveraging technology for waste sustainability: understanding the adoption of a new waste management system. Sustain Environ Res 33(1). https://doi.org/10.1186/s42834-023-00174-x

Wahlen S, Laamanen M (2015) Consumption, lifestyle and social movements. Int J Consum Stud 39(5):397–403. https://doi.org/10.1111/ijcs.12237

Wang J, Azam W (2024) Natural resource scarcity, fossil fuel energy consumption, and total greenhouse gas emissions in top emitting countries. Geosci Front 15(2):101757. https://doi.org/10.1016/J.GSF.2023.101757

World Bank Group (2018) What a waste 2.0: a global snapshot of solid waste management to 2050. Washington, D.C. https://doi.org/10.1596/978-1-4648-1329-0

World Bank (2022) Squaring the circle: policies from Europe's circular economy transition. https://www.worldbank.org/en/region/eca/publication/squaring-circle-europe-circular-economy-transition

Yu H, Zahidi I, Fai CM, Liang D, Madsen DØ (2024) Elevating community well-being in mining areas: the proposal of the mining area sustainability index (MASI). Environ Sci Eur 36(1). https://doi.org/10.1186/s12302-024-00895-9

Zavos S, Lehtokunnas T, Pyyhtinen O (2024) The (missing) social aspect of the circular economy: a review of social scientific articles. Sustain Earth Rev 7(1). https://doi.org/10.1186/S42055-024-00083-W

Chapter 5
Transforming Organizational Culture for the Circular Economy

Abstract This chapter explores the critical role of organizational culture in enabling the transition toward a circular economy, with a particular focus on Sustainable Development Goal 12 (SDG 12). It examines how cultural values, leadership styles, and internal education strategies shape the adoption of circular practices across diverse organizational contexts. Drawing on theoretical frameworks such as Stakeholder Theory, Open Systems Theory, and Contingency Theory, the chapter highlights the importance of systemic thinking, inclusive leadership, and generational engagement—especially among Millennials and Gen Z—in fostering sustainability-oriented cultures. The chapter emphasizes that circular transformation is not solely a technical endeavor; it requires a cultural shift that redefines organizational purpose, values, and behaviors. Leadership is presented as a key driver of change, with participative and transformational styles shown to enhance innovation, employee engagement, and alignment with circular principles. The chapter also discusses the influence of generational values on organizational sustainability, noting how younger cohorts prioritize purpose, ethics, and digital fluency—traits that support circular innovation but also pose challenges for long-term commitment. Organizational education is framed as a strategic enabler of circularity, focusing on the development of transversal competencies such as systems thinking, creativity, digital literacy, and ethical responsibility. These skills are essential for embedding circular practices into daily operations and decision-making. The chapter concludes by advocating for a holistic approach that integrates cultural, strategic, and educational dimensions, providing a robust foundation for aligning organizational behavior with circular economy principles and long-term value creation.

Keywords Organizational culture · Leadership styles · Open systems theory · Stakeholder engagement · Circular competencies · Generational values · Millennials · Gen Z

L. Velazquez, *Circular Economy and Sustainable Consumption in the Context of SDG 12*, Strategies for Sustainability, https://doi.org/10.1007/978-3-032-15598-6_5

5.1 Introduction

In today's business landscape—including not just industrial sectors but also services, technology, and finance—organizations face increasingly complex, volatile, and interconnected expectations that challenge traditional views of the purpose of the firm. A company is no longer just a vehicle for maximizing profits and productivity; it is also expected to take on broader social and environmental responsibilities. This expanded role reflects a shift in stakeholder expectations and regulatory frameworks, where legitimacy and long-term value creation depend not only on financial performance but also on ethical conduct, legal compliance, and contributions to societal well-being (Kurznack et al. 2021). As argue Salvioni and Almici (2020), this transformation requires a deep cultural shift toward sustainability, where corporate values and behaviors are reoriented to support circular economy principles and integrated value creation across economic, social, and environmental dimensions.

People are the most valuable asset of any organization, and their engagement is deeply shaped by the cultural environment in which they operate. When organizational culture is grounded in sustainability and circular economy principles, it becomes a powerful driver of change. It encourages individuals to rethink resource use, adopt regenerative practices, and align their actions with a shared vision of long-term well-being. A culture that values circularity not only fosters internal cohesion—it also empowers people to lead innovation and contribute meaningfully to a future where economic growth is balanced with environmental and social responsibility.

The transformation of the industrial mindset toward sustainability requires a deeper understanding of how organizational culture shapes strategic decision-making. As highlighted by Cordoba Maestre and García Samper (2025), different cultural archetypes—hierarchical, market-oriented, adhocratic, and clan—exert distinct influences on sustainability priorities. Hierarchical cultures tend to emphasize economic sustainability through control and efficiency, while market cultures focus on environmental outcomes driven by competitiveness. In contrast, adhocratic cultures, characterized by flexibility and innovation, are more conducive to holistic sustainability approaches, including circular economy practices.

This transformation is effectively illustrated by the evolution of Stakeholder Theory, which has expanded beyond its normative origins into a multidimensional framework that explains how organizations engage with diverse stakeholder groups. Contemporary interpretations emphasize co-creation of value, ethical governance, and strategic alignment with societal and environmental priorities—principles that are foundational to circular economy transitions (Mahajan et al. 2023). Contemporary interpretations emphasize mutual value creation, where stakeholders are not only recipients but also co-creators of value in joint processes of exchange (Freudenreich et al. 2020). Moreover, stakeholder theory has been applied to address sustainability challenges, ethical governance, and strategic alignment with societal values—including frameworks that integrate circular economy principles and stakeholder dialogue to foster collaborative value creation and long-term legitimacy (Awa et al. 2024).

To better understand this multidimensional role of the firm, it is useful to revisit foundational frameworks that conceptualize corporate responsibility beyond economic measures. One of the most influential models in this regard is the Pyramid of Corporate Social Responsibility (Carroll 1991), which outlines four interconnected dimensions: economic, legal, ethical, and philanthropic responsibilities. He argued that economic performance is fundamental but must be pursued alongside legal compliance, ethical conduct, and voluntary contributions to societal well-being. This framework suggests that profitability and legitimacy are not mutually exclusive but rather interdependent in a world where stakeholders increasingly demand transparency, fairness, and sustainability.

Adopting circular economy principles provides firms with a practical framework to align environmental responsibility with strategic performance. Practices such as eco-design, reverse logistics, and product life extension enable the reduction of waste, the recovery of valuable materials, and the minimization of virgin resource use (Al-Najjar et al. 2025; Hong et al. 2024). These strategies not only improve regulatory compliance and operational efficiency but also strengthen stakeholder relationships and foster brand loyalty (Mazzucchelli et al. 2022). Furthermore, Singh and Mishra (2024) emphasize that integrating circular economy with corporate social responsibility drives innovation, enhances economic resilience, and contributes directly to the achievement of the United Nations Sustainable Development Goals. Santiago et al. (2025) expand on this by positioning Corporate Social Responsibility (CSR) and circular practices as interdependent pillars within organizational culture, where waste is reframed as a resource and lifecycle thinking becomes central to decision-making. Their review highlights that organizations embracing cradle-to-cradle design, stakeholder engagement, and transparent governance not only reduce environmental impact but also cultivate long-term value creation aligned with SDG 12.

While many circular initiatives are supported by technological innovations, the transition to a circular economy is not solely a technical effort. Often, technologies serve as catalysts by improving resource efficiency, increasing traceability, and aiding decision-making processes (Giordano et al. 2024). For small- and medium-sized enterprises, reverse logistics systems supported by digital platforms and external partnerships are crucial for reintegrating materials into production cycles, despite financial and infrastructural challenges (Olipp et al. 2025). Additionally, digital technologies such as the Internet of Things, artificial intelligence, and blockchain have been shown to operationalize circular strategies across the value chain—optimizing supply chains, extending product life, and enabling closed-loop systems (Cagno et al. 2021). However, the success of these strategies relies not only on technological adoption but also on how organizations think, operate, and engage their people.

Transitioning to circularity must be rooted in organizational culture, guided by core values in daily decisions, and reflected in the systems that influence collective behaviors—going beyond engineering departments or sustainability teams to include all levels of the organization. Dos Santos Ferreira et al. (2024) emphasize that this transformation requires shifting from linear governance models to open systems thinking, enabling adaptive learning and systemic change aligned with circular principles. Enes and Durães (2025) highlight that purposeful leadership and corporate awareness foster emotional engagement and trust, which are essential for cultivating a culture of sustainability aligned with the goals of the 2030 Agenda. Barboza et al. (2022) illustrate that values such as adaptability, collaboration, transparency, systemic thinking, and empathy are foundational for embedding circularity into organizational behavior and decision-making. Muhammad Rehman and Ali Alqahtani (2025) provide empirical evidence that organizational culture enhances the effectiveness of circular economy drivers, reinforcing the need for cohesive cultural frameworks that support sustainable practices across diverse operational contexts.

Within this evolving cultural landscape, human capital plays a critical role in implementing circular principles across the organization. It is not only the technical expertise of individuals that matters but also their ability to internalize and embody the values that drive systemic change. Strategic human resource management practices—such as sustainability-focused training, adaptive learning models, and personalized development plans—are essential for equipping employees with green skills and fostering long-term commitment to circular goals (Teixeira 2025).

Additionally, Srisathan et al. (2020) suggest that employee empowerment and a positive organizational climate—marked by trust, autonomy, and shared values—are vital for encouraging innovative behaviors and supporting lasting transformation. Furthermore, green human capital directly contributes to sustainable production and the circular economy by helping firms develop dynamic capabilities that align environmental goals with operational performance (Cheng et al. 2023). Empirical evidence also indicates that green human resource management practices—particularly in recruiting and employee engagement—positively impact circular performance, although training and reward systems in this area remain underdeveloped, underscoring the need for targeted capacity-building efforts (Marrucci et al. 2021).

Organizational culture can be intentionally shaped to support circular economy transitions aligned with SDG 12. Sustainability-oriented cultures require shared assumptions about the interdependence between business and society; the imbalance among economic, social, and environmental domains; and the need to restore equilibrium. When these assumptions are embedded in a clear vision and values, and consistently communicated, they foster emotional commitment among organizational members. The following sections explore how leadership can activate cultural change, how organizational models shape the diffusion of circular values, and how targeted education and empowerment strategies build the competencies needed to sustain circularity across diverse contexts.

5.2 Leadership for Circular Transformation

As organizations transition from linear to circular strategies, leadership must adapt to promote collaboration, inclusivity, and sustainability-focused behaviors at all levels. Recent empirical research by Kantabutra (2021) introduces the Integrated Sustainability Organizational Culture Model, which positions leadership as a central driver of corporate sustainability. The model emphasizes that sustainability-oriented leadership must begin with a clearly articulated and emotionally resonant vision—one that is abstract, future-oriented, and inspiring. When such a vision is consistently communicated alongside core sustainability values, it fosters emotional commitment among organizational members. This commitment, in turn, activates five key sustainability practices: perseverance, moderation, resilience development, geosocial development, and knowledge-sharing. Complementing this perspective, Fan et al. (2021) demonstrate that leadership style significantly influences organizational innovation, especially in the context of Industry 4.0 and sustainability transitions. Their findings suggest that while leadership remains a key factor, its impact is most effective when aligned with bottom-up cultural dynamics and customer orientation. Moreover, they emphasize that enabling leadership should not only guide strategic direction but also foster knowledge integration capabilities among employees—an essential condition for translating sustainability goals into innovative organizational practices.

Inclusive leadership plays a pivotal role in this transformation. Ly (2025) defines it as a style characterized by openness, accessibility, and shared decision-making, which enhances the effectiveness of green human resource management practices and supports circular economy performance. Building on this, recent studies emphasize that leadership motivation significantly influences the depth and scope of circular adoption. Howe and Lavy (2025) demonstrate that intrinsic motivation among company leaders correlates strongly with the implementation of more complex circular practices, such as circular design and employee education, while extrinsic motivation tends to drive only basic compliance-oriented actions. Sulich et al. (2021) contribute to this discussion by proposing a color-based taxonomy of management styles—ranging from directive (red) to self-managed and purpose-driven (teal)—each associated with distinct leadership profiles and strategic orientations. In this framework, transformational and participative leaders, aligned with orange and green styles, foster cultures of innovation, empowerment, and ecological awareness. These styles are particularly conducive to circular transformation, as they balance flexibility with strategic coherence and promote shared responsibility across organizational levels. Leadership effectiveness in circular contexts thus depends not only on managerial capabilities but also on the alignment between leadership style, organizational culture, and sustainability values.

This emphasis on autonomy, empowerment, and systemic thinking in leadership naturally extends to broader organizational design principles, particularly those of human-centered organizations. Human-centered organizations, as described by Townsend and Romme (2024), prioritize autonomy, psychological safety, and personal growth, recognizing individuals as whole persons with creative and self-directed potential. Ideally, the workplace is a space where employees find meaningful roles, fair treatment, and supportive relationships that enable them to thrive both professionally and personally. This complexity becomes even more pronounced in the context of the circular economy, where the International Labour Organization (ILO) emphasizes that decent work must be a foundational element of any transition. Circular activities—such as repair, reuse, and recycling—are often labor-intensive and, in many cases, carried out under informal or precarious conditions. While these activities contribute to environmental sustainability, they do not inherently guarantee fair wages, safe working environments, or social protections (ILO 2023). Therefore, human-centered organizations must go beyond inclusion and autonomy to actively ensure labor standards, occupational safety, and equitable access to training and reskilling opportunities.

The information technology (IT) sector, which includes companies that develop software, digital services, infrastructure, and computing solutions, offers a compelling example. Organizations in this sector that adopt circular economy principles tend to emphasize innovation, continuous product improvement, inclusive employment practices, and vocational training. These efforts align with the goals of Sustainable Development Goal 8 (SDG 8): Decent Work and Economic Growth, by fostering resilient, equitable, and future-oriented workplaces. However, without intentional leadership and policy frameworks, circular transitions may inadvertently reproduce existing inequalities—especially in industries with gendered cultures or uneven access to digital skills (Yücel Karamustafa et al. 2024).

The convergence of human-centered organizational design and circular economy principles offers a promising path toward sustainable and inclusive work environments. However, achieving this potential requires intentional efforts to align values, governance, and operational practices with the realities of diverse labor markets and their structural challenges. While such organizations promote inclusion and empowerment, managers still need to handle a wide range of employee attitudes, motivations, and behaviors that affect how work is approached. Organizations are made up of individuals with different beliefs, personal values, and lived experiences, which means that fostering a positive culture also involves negotiating with various types of workers and management styles.

Theory X and Theory Y, as presented by Kopelman (2023), exemplify two contrasting views of human nature that influence organizational culture: Theory X assumes employees dislike work and need control, while Theory Y sees them as intrinsically motivated and capable of self-direction. Beyond these, Theory Z integrates elements of Japanese and American management to foster long-term employment and trust-based governance. While Theories X and Y focus on managerial assumptions about human nature—ranging from control-oriented to empowerment-based views—Theory Z offers a shift in perspective by emphasizing the role of organizational culture in shaping employee behavior and commitment. As explained by Daft (2004), Theory Z is better understood as a management philosophy than a formal theory. Developed by William Ouchi, it integrates key aspects of Japanese and American organizational models to foster long-term employment, trust-based governance, and a holistic concern for employee well-being. These principles align with sustainability-oriented cultures by emphasizing stability, loyalty, and participatory decision-making—conditions that support the adoption of circular economy practices.

The diversity of employee mindsets—shaped by personal values, experiences, and attitudes—significantly influences how individuals engage with organizational goals, adapt to change, and interpret sustainability initiatives. Recognizing and embracing this diversity is vital for fostering a culture that supports circular economy principles and encourages inclusive transformation. Specifically, corporate sustainability—fundamental to circular economy models—may be perceived and valued differently by each member, leading to tensions between official messages and actual practices. Recognizing and adjusting to these differences is crucial for aligning circular economy goals with inclusive and effective workforce management. Within this nuanced context, the evolution of leadership theory becomes relevant. As Benmira and Agboola (2021) explain, leadership has advanced from trait-based and behavioral models to situational and systemic approaches.

Recent studies reinforce the importance of leadership styles that are not only adaptive but also focused on sustainability. Piwowar-Sulej and Iqbal (2023) demonstrate that transformational and sustainable leadership styles are especially effective in improving environmental and organizational performance, particularly when aligned with long-term stakeholder values. As Bashynska et al. (2024) further develop this idea by presenting a multidimensional framework for evaluating sustainable leadership in circular economy-driven innovation, emphasizing the integration of personal, managerial, and innovation-focused components. Their model indicates that leadership effectiveness depends not only on strategic direction but also on personal skills—such as values, thinking, and emotional intelligence—that influence innovation outcomes. Costa et al. (2023) for their part add that participatory leadership styles, especially transformational and democratic, have a positive effect on innovation when combined with strong human capital.

Fostering innovation, intellectual stimulation, and a shared vision are essential qualities for navigating complex and dynamic organizational environments. Transformational leadership embodies these traits and plays a pivotal role in guiding organizations toward sustainable change. As Bwalya (2015) highlights, transformational leaders inspire individuals to go beyond personal interests in pursuit of collective goals, while promoting creativity, empathy, and adaptability. This leadership style aligns closely with the principles of organization-centered management theories, emphasizing that sustainable change results not from isolated actions but through continuous, multi-level learning processes influenced by environmental and social interdependencies.

5.3 Generational Leadership for Circular Transformation

As Millennials and Generation Z increasingly make up the majority of the global workforce, understanding their values, motivations, and expectations is crucial for creating leadership models that effectively support circular transformation. These groups bring unique perspectives shaped by digital fluency, increased environmental consciousness, and a strong desire for purpose-driven work. For human resource managers and organizational leaders, this means developing inclusive and adaptable strategies that align with these generations' priorities—especially in contexts where long-term sustainability and innovation are key. According to the Gen Z and

Millennial Survey by Delloitte Global (2025), 89% of Gen Z and 92% of Millennials consider a sense of purpose vital to job satisfaction and well-being. Furthermore, 41% of Gen Z and 40% of Millennials have rejected potential employers over ethical concerns, and their environmental values influence employment choices—23% of Gen Z and 22% of Millennials have researched a company's environmental policies before accepting a job.

On the other hand, these cohorts are not primarily motivated by climbing the corporate ladder; only 6% of Gen Z identify reaching leadership positions as their main career goal. Instead, they seek continuous learning, personal growth, and meaningful contributions. While this orientation aligns well with the collaborative and values-based nature of circular economy initiatives, it also presents a challenge: projects aimed at circular transformation often require long-term commitment, strategic continuity, and sustained leadership engagement. A workforce that is disengaged from long-term organizational goals or prone to frequent job changes—31% of Gen Z and 17% of Millennials plan to leave their current employer within 2 years—may hinder the stability and perseverance needed to drive systemic change.

While Deloitte's 2025 Global Survey emphasizes the personal values and ethical motivations of Millennials and Gen Z, an updated World Economic Forum (2025) "Future of Jobs Report" provides a complementary perspective by highlighting the evolution of skills and workforce strategies that employers are prioritizing in response to global macrotrends. The report identifies AI and big data, networks and cybersecurity, and technological literacy as the three fastest growing skills projected through 2030, followed by creative thinking, resilience, flexibility, agility, and environmental stewardship—all of which align with the values of younger generations. Importantly, the report underscores that digital technologies are not only transforming business models but are also essential enablers of the green transition, including circular economy initiatives.

These insights reinforce the strategic relevance of Millennials and Gen Z as agents of change—not only because of their values and ethical motivations, but also due to their digital competencies, which are essential for implementing circular economy strategies. Across multiple sectors, digital tools such as smart energy grids, computational design workflows, and Building Information Modeling (BIM) systems (Munonye 2025), as well as Big Data analytics in small- and medium-sized enterprises (Almanza Junco et al. 2023), have proven to be key enablers of circular innovation. Without digitally skilled human capital, organizations risk falling short in their efforts to operationalize circular strategies, measure impact, and adapt to evolving sustainability standards.

At the same time, some scholars caution against over-relying on generational categories, arguing that the empirical evidence for meaningful differences is limited. Clements (2025) highlights that generational identities often carry a "social reality"—many individuals assume these categories are real and behave accordingly. This perception can shape workplace dynamics, including expectations around leadership and organizational culture. Esenyel (2024) builds on this by emphasizing that effective leadership in volatile, uncertain, complex, and ambiguous (VUCA) environments requires adaptive and heliotropic leadership styles—those that foster growth, positivity, and alignment with shared values. These approaches resonate with the expectations of Millennials and Gen Z, supporting inclusive, resilient, and digitally fluent organizational cultures. Rudolph et al. (2018) further argue that rather than relying on generational stereotypes, leadership development should adopt a lifespan perspective, recognizing individual trajectories and fostering intergenerational learning. This shift not only avoids reductive assumptions but also aligns with the broader goal of cultivating circular competencies across diverse age groups.

By integrating adaptive and inclusive leadership with a nuanced understanding of generational dynamics, managers can empower younger employees not only to engage meaningfully but also to lead transformative initiatives toward sustainability. This alignment between leadership style and generational values is essential for embedding circular principles into organizational culture and practice.

5.4 Organization-Centered Management Theories

Theories of leadership and management are deeply intertwined, particularly in the context of organizational transformation toward sustainability and circularity. While leadership theories focus on the behaviors, styles, and interpersonal dynamics that guide individuals and teams, organization-centered management theories—such as Open Systems Theory, Contingency Theory, and Institutional Theory—provide the structural and contextual frameworks within which leadership operates. These theories explore how organizations adapt, evolve, and interact with their environments, emphasizing responsiveness, systemic alignment, and institutional legitimacy.

This perspective helps explain how organizations respond to external pressures via feedback mechanisms, portraying them not as static or isolated structures, but as permeable systems whose survival, adaptation, and evolution depend on their environmental relationships—unlike closed systems. Gearty et al. (2015) expand this understanding by framing sustainable transformation as a form of deep systems change, where learning unfolds across interconnected actors and institutions. Their work emphasizes that addressing sustainability challenges requires iterative, multilevel learning processes that transcend organizational boundaries and engage diverse stakeholders in adaptive, experiential reflection—an approach that resonates strongly with the open systems logic.

Contingency theory offers another valuable framework for understanding how organizations can adapt their sustainability strategies to specific internal and external conditions. As Yusuf et al. (2023) demonstrate, contingency theory helps explain how variables such as corporate sustainable support policies and incentive structures significantly influence the adoption of sustainable innovation in manufacturing firms. Their empirical findings show that when companies align these organizational factors with contextual realities, they are more likely to foster innovation that supports long-term environmental and social goals. Complementing this, Morais and Barbieri (2022) apply contingency theory to supply chain social sustainability, revealing how factors like power dynamics, material criticality, and supplier dependence shape governance mechanisms for addressing social issues. Together, these studies highlight the theory's relevance for advancing both sustainability and circular economy practices by promoting flexible, context-sensitive approaches to organizational and supply chain transformation.

As a third theoretical perspective, Institutional Theory shifts the focus toward the external structures and cultural expectations that shape organizational behavior over time, offering a distinct explanation for how sustainability and circular economy practices emerge and gain legitimacy. Unlike contingency theory, which highlights strategic alignment with contextual factors, institutional theory centers on how norms, regulations, and cultural expectations influence organizational behavior over time. Dodds and Smith (2025) demonstrate how coercive, mimetic, and normative pressures collectively promote sustainability transitions in tourism-dependent communities. Their longitudinal study of Tofino, Canada, shows how regulatory mandates (coercive), imitation of best practices (mimetic), and professional norms (normative) interact to drive systemic change toward sustainability. Likewise, Glover et al. (2014) apply institutional theory to the dairy supply chain, revealing how dominant actors like supermarkets exert coercive pressure on smaller suppliers to adopt energy-efficient practices, while normative and mimetic forces strengthen legitimacy and visibility of sustainable actions.

In corporate reporting, Tudor-Tiron and Dragu (2014) demonstrate that integrated reporting—rooted in sustainability practices—is shaped by institutional factors like legal systems, highlighting how environmental efforts gain legitimacy through structured disclosure. Recent reviews Galleli and Amaral (2025) further stress that sustainability and circular initiatives develop through the interaction of institutional pressures and actor responses across multiple levels. This perspective underscores how organizations not only adapt to external expectations but also help reshape institutional norms over time. In the context of the circular economy, institutional theory provides a useful perspective to understand how circularity practices—such as integrated reporting, resource efficiency, and regenerative models—become legitimate and embedded within organizational processes, indicating a shift from mere compliance to organizational transformation.

5.5 Evolution of Management Theories Toward Circular Economy

Although existing organizational theories offer valuable insights and frameworks that can support a transition toward sustainability, the shift has yet to materialize in a meaningful and widespread way. As Ketprapakorn and Kantabutra (2022) argue, most current models of sustainability-oriented organizational culture remain incomplete, lacking key elements such as shared assumptions, values, and artifacts that are essential for guiding transformative change. Their integrated theory of sustainability organizational culture highlights the need for a dynamic, systems-based approach that recognizes organizations as open systems embedded in broader institutional and ecological contexts.

Scholars have explored the underlying reasons for this theoretical and practical gap, pointing to structural, epistemological, and cultural limitations. Wasieleski et al. (2021) suggest that dominant management theories—grounded in economic rationality and individualism—struggle to address the complexity of the Anthropocene. Rather than dismissing these theories outright, they call for a paradigm shift that integrates knowledge from the natural sciences and embraces a holistic view of the planet as an interconnected system. In a similar vein, Callaghan and Mitchell (2024) critique the rigidity of classical frameworks such as Burrell and Morgan's paradigm model. They propose a post-paradigmatic epistemology known as ecosystemic panarchy, which emphasizes multi-level, adaptive, and non-linear thinking. This approach integrates ecological imperatives into organizational theory development, encouraging a more holistic and responsive understanding of sustainability challenges. Building on these perspectives, Carbal Herrera et al. (2024) emphasize the potential of newer organizational approaches—such as the natural resource-based view, stakeholder theory, and strategy microfoundations—for integrating sustainability into practice. However, they warn that these frameworks often remain limited by an instrumental logic centered on short-term profitability and efficiency. This underscores the broader need for a paradigm shift toward organizational theories that are ethically grounded and ecologically responsive, capable of tackling sustainability challenges through comprehensive cultural and systemic transformation.

The organizational change toward a circular economy involves not only technical transformations in products or business models but also requires a profound and simultaneous reconfiguration of human and organizational aspects. This shift demands changes in organizational culture, shared values, internal structures, and innovation processes, as well as collaboration with external stakeholders. As highlighted by Graessler et al. (2024), the success of Organizational Change toward Circular Economy depends on the holistic integration of technical, human, and organizational dimensions, since addressing them in isolation can undermine the effectiveness of change efforts and increase internal resistance. This sociotechnical perspective underscores the need to align technological capabilities with social and organizational practices to enable a sustainable and lasting circular transition.

Organizational culture can be understood as the set of shared norms, values, attitudes, and expectations that guide the behavior of members within an organization. Although not directly observable, it becomes evident through its effects on internal relationships, decision-making processes, and the achievement of institutional goals (Sánchez Castro 2021). This foundational view has been enriched by various theoretical perspectives. Barón Ballesteros (2024) notes that organizational culture has been conceptualized as a model, symbol, practice, or system, and despite definitional differences, scholars agree that it involves a framework of beliefs and norms that shape both symbolic and operational environments within intelligent organizations. Mena Méndez (2024) adds that organizational culture has evolved in response to globalization, technological advancement, and market demands, describing it as a system of corporate values, socialization mechanisms, cultural change processes, and mediations that influence institutional development. Similarly, Llanos Encalada (2016) characterizes it as the "personality of the organization," a set of traits that confer identity and distinguish it from other entities. This personality is reflected in both internal and external relationships and serves as a strategic axis for talent management.

In this same vein, Huertas López et al. (2020) emphasize that organizational culture is not merely a symbolic construct but a system of institutional capabilities that must remain in balance with environmental demands. From their perspective, management models should incorporate culture as a structural axis that connects organizational processes, policies, and human talent competencies. This view reinforces the notion that culture is a strategic dimension—essential for operational sustainability and institutional adaptability. In the context of circular economy transitions, this perspective aligns with the need for resilient organizations capable of reconfiguring their internal logic to support regenerative practices, long-term value creation, and systemic innovation. By embedding cultural principles that foster flexibility, collaboration, and ecological awareness, organizations can better respond to uncertainty and complexity—core challenges in circular design and implementation.

According to Wang et al. (2023), organizational culture can be classified into four distinct types based on the Competing Values Framework: clan, adhocracy, market, and hierarchy cultures. Each type represents a different focus and set of values. Clan culture emphasizes teamwork, loyalty, and interpersonal bonds, creating a family-like environment. Adhocracy culture promotes innovation, flexibility, and risk-taking, fostering creativity and adaptability. Market culture centers on competitiveness, goal achievement, and productivity, driven by external results. Lastly, a hierarchical culture values stability, control, and formal procedures, relying on structured processes and clearly defined responsibilities. These cultural types coexist in varying degrees within organizations and influence both individual and organizational performance.

While all organizational culture types have their own advantages and limitations, transitioning toward a circular economy tends to thrive in environments that embrace cultural and creative practices aligned with the strategies of Refuse, Rethink, Reduce, Reuse, Repair, Refurbish, Remanufacture, Repurpose, and Recycle, Pedrini and Tartari (2023). Cultures with an adhocratic orientation naturally support experimentation and openness to change—qualities essential for reimagining resource use and redesigning systems. This is particularly relevant given that circular innovation—both technological and non-technological—requires organizations to transform not only products and processes, but also internal practices and mindsets. Such transformation is more likely to succeed when driven by entrepreneurial orientation and supported by circular human resource management (Castro-Lopez et al. 2025).

Empirical evidence from (Butt et al. 2024) reinforces this perspective by demonstrating that organizational culture significantly influences sustainable performance, both directly and indirectly, through the mediation of eco-innovation and circular economy practices. Their study, based on 200 manufacturing firms in Pakistan, found that cultures fostering ethical leadership, employee engagement, and strategic alignment with sustainability are more likely to adopt circular practices and achieve improved environmental, social, and economic outcomes. In line with this, broader research in the hospitality sector, Elshaer et al. (2024) support the notion that the integration of circular economy principles within organizations is most effective when aligned with environmentally conscious human resource practices and a supportive cultural environment.

Additionally, Muzondo et al. (2025) emphasize that the successful adoption of circular models in resource-constrained environments depends not only on cultural alignment but also on the availability and mobilization of internal capabilities—such as technical knowledge, infrastructure, and collaborative networks. Complementing these insights, Sadaoui et al. (2025) argue that while organizational culture may not always mediate the relationship between technological innovation and circular economy adoption, it acts as a stable enabler that provides interpretive coherence and normative support throughout the implementation process.

As Opferkuch et al. (2021) argue, embedding sustainability into organizational strategy requires internal commitment and cross-functional collaboration, conditions that clan cultures naturally foster through shared values and participatory decision-making. Similarly, Riemer et al. (2025) emphasize that the development of a Culture of Sustainability is an emergent and iterative process, where leadership and employee engagement converge to integrate sustainability into the organizational identity. These approaches align with circular principles by cultivating environments where value-driven innovation, adaptability, and long-term thinking are embedded in everyday practices. Ultimately, what matters most is not selecting a single cultural model but cultivating a balance that reflects the organization's values and sincere commitment to change. When leadership encourages autonomy, purpose, and shared responsibility, the journey toward circularity becomes not only achievable but also meaningful.

5.6 Organizational Education as a Circularity Driver

As organizations embrace circular economy principles, the development of internal competencies becomes a strategic imperative. Organizational education is not merely a support function—it is a transformative lever that enables cultural alignment, systemic thinking, and innovation across all levels. In the context of SDG 12, fostering circular competencies through targeted learning strategies is essential for embedding sustainability into everyday practices and decision-making.

Transitioning toward circularity requires organizations to invest not only in technological infrastructure but also in the continuous development of their human capital. As Rizos et al. (2016) and Ogunmakinde et al. (2025) emphasize, circular competencies must include not only technical skills but also transversal capabilities such as systems thinking, ethical reasoning, and innovation. These competencies enable employees to contribute meaningfully to regenerative practices from their specific roles and contexts.

The circular economy, by its very nature, demands the integration of knowledge, skills, and approaches from diverse sectors and disciplines. In today's workforce, this translates into a call for collaboration among governments, educational institutions, businesses, and civil society to implement systemic changes that optimize the use of material and energy resources. These solutions must be applied coherently across multiple scales—micro (product and user), meso (infrastructure and community), and macro (city and territory)—to effectively address contemporary environmental, social, and economic challenges (Drofenik et al. 2025; Ghisellini et al. 2023; Xu et al. 2020).

For professionals already engaged in the labor market, many of the principles of the circular economy may resonate with their practical experience, even if these concepts were not explicitly addressed during their formal education. Therefore, the progressive development of key transversal competencies—such as systems thinking, creativity and innovation, digital literacy, ethical decision-making, and adaptability to changing contexts—should be supported through continuous learning strategies that recognize prior experiential knowledge and offer relevant, accessible opportunities for upskilling (Mogena Medina et al. 2023; Soto Kiewit et al. 2023).

5.6.1 Systems Thinking

In professional environments, systems thinking is increasingly recognized as a strategic skill for navigating complexity. Rather than focusing on isolated problems, this approach encourages professionals to identify patterns, feedback loops, and root causes within interconnected systems. It enables a deeper understanding of how decisions in one area can ripple across others—social, economic, and ecological alike. For workers already immersed in operational realities, systems thinking can be cultivated through reflective practice, cross-functional collaboration, and exposure to real-world problem-solving scenarios (De Angelis and Ianulardo 2025; Ndaruhutse et al. 2019).

5.6.2 Creativity and Innovation

Creativity in the workplace is not limited to artistic expression—it is a practical tool for rethinking processes, products, and services. Innovation, especially in the context of circular economy, requires professionals to challenge conventional models and propose alternatives that are both sustainable and feasible. This involves integrating diverse perspectives, experimenting with new approaches, and adapting solutions to specific organizational contexts. Encouraging creativity among employees can lead to eco-innovative outcomes that reduce resource dependency and enhance resilience (Bucur 2024; Park et al. 2021).

5.6.3 Digital Competence

Digital tools are essential for enabling circular practices such as resource tracking, predictive maintenance, and smart logistics. For professionals, digital competence goes beyond basic Information Technology (IT) skills—it includes the ability to leverage technologies like data analytics, Internet of Things (IoT), and Artificial Intelligence (AI) to support sustainable decision-making. Upskilling in digital areas empowers workers to participate in circular value chains and contribute to data-driven sustainability strategies. Organizations must invest in accessible training programs that bridge digital gaps and promote inclusive technological adoption (Jensen 2025; Karanfiloğlu 2025).

5.6.4 Ethical and Environmental Responsibility

Ethical awareness and environmental responsibility are no longer optional in professional settings—they are central to sustainable performance. Workers are increasingly expected to consider the long-term impacts of their actions, not only on profitability but also on people and the planet. This involves making choices that align with values such as fairness, transparency, and ecological stewardship. Cultivating these competencies requires ongoing dialogue, ethical leadership, and a workplace culture that prioritizes sustainability over short-term gains (Lozano 2015; United Nations Educational, Scientific, and Cultural Organization [UNESCO] 2020).

In addition to general transversal skills, professional development in the circular economy requires specialized competencies that enable practitioners to apply circular principles within concrete technical, operational, and organizational contexts. Because circularity spans a wide range of sectors, scales, and strategic approaches, the set of relevant skills is extensive—and potentially limitless—depending on the area of application, level of intervention, and specific goals. As Renfors (2024) highlights, professionals must be equipped to translate circular concepts into actionable practices. These competencies are essential, but not limited to: efficient resource management, product design based on circular criteria (such as eco-design and life-cycle assessment), development of circular business models, understanding of regulatory and market dynamics, and inter-organizational collaboration aimed at reducing environmental impacts.

Although many workers have completed formal education—whether at the upper secondary or tertiary level—it is likely that they have developed competencies such as systems thinking, creativity, digital literacy, and ethical decision-making through work experience, informal learning, or self-directed study. However, a significant portion of employees may not have encountered these key competencies during their academic training, which poses challenges for transitioning toward circular models. In this context, it is essential that continuous training strategies acknowledge prior experiential knowledge while offering accessible, context-sensitive, and flexible programs that help close knowledge gaps and foster an organizational culture aligned with sustainability principles. Non-formal and informal learning pathways—such as free online courses, virtual diplomas, and community-based workshops—have proven to be effective tools for democratizing access to circular economy knowledge and strengthening the capabilities of workers across all levels (Blomsma and Brennan 2017).

To address existing gaps and foster the continuous development of circular competencies among professionals, organizations must adopt diverse and inclusive educational strategies. Methods such as Massive Open Online Courses (MOOCs), hybrid workshops, digital simulations, and communities of practice have proven effective for lifelong learning and workplace training. Their success, however, depends on the design of active learning experiences that transcend theoretical instruction. Approaches like problem-based, project-based, and challenge-based learning promote critical, systemic, and multidisciplinary thinking—skills essential for circular innovation (Giannoccaro et al. 2021). MOOCs, in particular, offer flexible and affordable access to knowledge, enabling professionals to update their skills and engage with sustainability-oriented content at their own pace (Atiaja Atiaja and García Martínez 2020). When tailored to circular economy principles, these platforms can simultaneously cultivate digital literacy and ecological awareness (Loste et al. 2017). By embedding such educational practices into organizational culture, companies can empower their workforce to lead regenerative change, ensuring that circularity becomes a lived reality across all levels of decision-making and operations.

5.7 Conclusion

The transition toward a circular economy is not merely a technical or operational challenge—it is a cultural transformation that requires organizations to rethink their values, leadership models, and educational strategies. As demonstrated throughout this chapter, organizational culture plays a pivotal role in enabling circularity by shaping behaviors, decision-making processes, and stakeholder relationships. Cultures grounded in sustainability foster emotional commitment, innovation, and systemic thinking, all of which are essential for achieving SDG 12.

Leadership emerges as a key driver of this transformation, particularly when it is inclusive, purpose-driven, and aligned with the values of emerging generations. Millennials and Gen Z bring ethical awareness, digital fluency, and a desire for meaningful work, positioning them as strategic allies in the pursuit of regenerative business models. However, their engagement must be supported by organizational structures that promote autonomy, trust, and long-term development.

Theoretical frameworks such as Open Systems Theory, Contingency Theory, and Institutional Theory offer valuable insights into how organizations can adapt to complex sustainability demands. These models underscore the importance of responsiveness, contextual alignment, and legitimacy in embedding circular principles into organizational logic.

Finally, organizational education is a cornerstone for building circular competencies across all levels. By investing in active, inclusive, and context-sensitive learning strategies, organizations can equip their workforce with the transversal and technical skills needed to operationalize circular economy practices. This holistic approach—integrating culture, leadership, theory, and education—provides a robust foundation for sustainable transformation and long-term value creation.

References

Almanza Junco CA, Pulido Ramirez MP, Gaitán Angulo M, Gómez-Caicedo MI, Mercado Suárez ÁL (2023) Factors for the implementation of the circular economy in Big Data environments in service companies in post pandemic times of COVID-19: the case of Colombia. Front Big Data 6. https://doi.org/10.3389/FDATA.2023.1156780/FULL

Al-Najjar B, Salama A, Abed A (2025) From compliance to commitment: examining the influence of corporate governance and social performance on firms' circular economy initiatives. J Bus Ethics.https://doi.org/10.1007/S10551-025-06111-9

Atiaja Atiaja LN, García Martínez A (2020) Los MOOC: Una alternativa para la formación continua. Rev Sci 5(18):120–136. https://doi.org/10.29394/SCIENTIFIC.ISSN.2542-2987.2020.5.18.6.120-136

Awa HO, Etim W, Ogbonda E (2024) Stakeholders, stakeholder theory and corporate social responsibility (CSR). Int J Corp Soc Responsib 9(1). https://doi.org/10.1186/S40991-024-00094-Y

Barboza LL, Bertassini AC, Gerolamo MC, Ometto AR (2022) Organizational values as enablers for the circular economy and sustainability. Revista de Administração de Empresas 62(5). https://doi.org/10.1590/S0034-759020220509

Barón Ballesteros R (2024) Cultura Organizacional: Una Reflexión Teórica Hacia la Construcción de Organizaciones Inteligentes. Ciencia Latina Revista Cientìica Miltidisciplinar 8(2). https://doi.org/10.37811/cl_rcm.v8i2.10585

Bashynska I, Malynovska Y, Kolinko N, Bielialov T, Järvis M, Kovalska K, Saiensus M (2024) Performance assessment of sustainable leadership of enterprise's circular economy-driven innovative activities. Sustainability 16(2). https://doi.org/10.3390/SU16020558

Benmira S, Agboola M (2021) Evolution of leadership theory. BMJ Leader 5(1):3–5. https://doi.org/10.1136/LEADER-2020-000296

Blomsma F, Brennan G (2017) The emergence of circular economy: a new framing around prolonging resource productivity. J Ind Ecol 21(3):603–614. https://doi.org/10.1111/jiec.12603

Bucur LM (2024) Eco-innovation and higher education in the European Union: exploring the green interconnection. Stud Bus Econ 19(1):46–60. https://doi.org/10.2478/SBE-2024-0003

Butt AS, Ali I, Govindan K (2024) The role of reverse logistics in a circular economy for achieving sustainable development goals: a multiple case study of retail firms. Prod Plan Control 35(12):1490–1502. https://doi.org/10.1080/09537287.2023.2197851

Bwalya A (2015) Leadership styles. Leadership in Sport (August):34–47. https://doi.org/10.1108/01437730510575570

Cagno E, Neri A, Negri M, Bassani CA, Lampertico T (2021) The role of digital technologies in operationalizing the circular economy transition: a systematic literature review. Appl Sci 11(8). https://doi.org/10.3390/APP11083328

Callaghan CW, Mitchell A (2024) Environmental sustainability and management theory development: post-paradigm insights from the Anthropocene. Eur Manag Rev 21(3):520–532. https://doi.org/10.1111/emre.12567

Carbal Herrera AE, Luna Amador JA, Caraballo Payares AM (2024) Organizational theory and business sustainability: a theoretical analysis. Revista de Ciencias Sociales 30(ESPECIAL 10):531–550. https://doi.org/10.31876/rcs.v30i.42858

Carroll AB (1991) The pyramid of corporate social responsibility: toward the moral management of organizational stakeholders. Bus Horiz 34(4):39–48. https://doi.org/10.1016/0007-6813(91)90005-G

Castro-Lopez A, Entrialgo M, Liao CT, Santos-Vijande ML (2025) Towards circular economy through innovation: the role of entrepreneurial orientation and human resource management. Int Entrep Manag J 21(1). https://doi.org/10.1007/s11365-024-01032-x

Cheng Y, Masukujjaman M, Sobhani FA, Hamayun M, Alam SS (2023) Green logistics, green human capital, and circular economy: the mediating role of sustainable production. Sustainability 15(2). https://doi.org/10.3390/SU15021045

Clements AJ (2025) A critical review of research on work-related differences in generational cohorts. Occup Psychol Outlook 4(1):27–40. https://doi.org/10.53841/BPSOPO.2025.4.1.27

Cordoba Maestre JP, García Samper M (2025) The influence of organizational culture on business process management: an update of a systematic literature review. Int J Environ Sci 11(23):759–767. https://doi.org/10.5753/sbsi.2025.246632

Costa J, Pádua M, Moreira AC (2023) Leadership styles and innovation management: what is the role of human capital? Adm Sci 13(2). https://doi.org/10.3390/ADMSCI13020047

Daft RL (2004) Theory Z: opening the corporate door for participative managment. Acad Manag Exec 18(4):117–121. https://doi.org/10.5465/AME.2004.15268709

de Faria Enes R, Durães RSS (2025) Exploring the circular economy: an analysis of leadership, corporate awareness and organizational culture. Int J Prof Bus Rev 10(1):e05197. https://doi.org/10.26668/BUSINESSREVIEW/2025.V10I1.5197

De Angelis R, Ianulardo G (2025) Walking the talk on education for sustainable development: the value of circular economy education in the age of complexity. Circ Econ Sustain. https://doi.org/10.1007/S43615-025-00577-5

Delloitte Global (2025) Gen Z and millenial survey. https://www.deloitte.com/global/en/issues/work/genz-millennial-survey.html

Dodds R, Smith W (2025) Institutional theory—assessing longitudinal change towards sustainability in Tofino, Canada. J Destin Mark Manag 37(March). https://doi.org/10.1016/j.jdmm.2025.101001

Dos Santos Ferreira C, Gomes G, Castillo-Ospina DA, Bertassini AC, Zaguetto C, Feltrin N et al (2024) Organizational practices, values, and mindsets as a basis for circular economy transition. In: A systemic transition to circular economy, pp 85–110. https://doi.org/10.1007/978-3-031-55036-2

Drofenik J, Seljak T, Novak Pintarič Z (2025) A multi-level approach to circular economy progress: linking national targets with corporate implementation. J Clean Prod 493:144902. https://doi.org/10.1016/J.JCLEPRO.2025.144902

Elshaer IA, Azazz AMS, Kooli C, Alqasa KMA, Afaneh J, Fathy EA et al (2024) Resilience for sustainability: the synergistic role of green human resources management, circular economy, and green organizational culture in the hotel industry. Adm Sci 14(11). https://doi.org/10.3390/ADMSCI14110297

Esenyel V (2024) Evolving leadership theories: integrating contemporary theories for VUCA realities. Adm Sci 14(11). https://doi.org/10.3390/ADMSCI14110270

Fan YJ, Liu SF, Luh DB, Teng PS (2021) Corporate sustainability: impact factors on organizational innovation in the industrial area. Sustainability 13(4):1–24. https://doi.org/10.3390/SU13041979

Freudenreich B, Lüdeke-Freund F, Schaltegger S (2020) A stakeholder theory perspective on business models: value creation for sustainability. J Bus Ethics 166(1):3–18. https://doi.org/10.1007/S10551-019-04112-Z

Galleli B, Amaral L (2025) Bridging institutional theory and social and environmental efforts in management: a review and research Agenda. J Manag XX. https://doi.org/10.1177/01492063251322429

Gearty MR, Bradbury-Huang H, Reason P (2015) Learning history in an open system: Creating histories for sustainable futures. Manag Learn 46(1):44–66. https://doi.org/10.1177/1350507613501735

Ghisellini P, Patil RA, van Langen SK, Ramakrishna S (2023) Circularity at Meso level: a sector perspective. In: Circularity assessment: macro to nano: accountability towards sustainability, pp 57–73.https://doi.org/10.1007/978-981-19-9700-6_4/FIGURES/3

Giannoccaro I, Ceccarelli G, Fraccascia L (2021) Features of the higher education for the circular economy: the case of Italy. Sustainability (Switzerland) 13(20):1–25. https://doi.org/10.3390/su132011338

Giordano V, Castagnoli A, Pecorini I, Chiarello F (2024) Identifying technologies in circular economy paradigm through text mining on scientific literature. PLoS ONE 19(12). https://doi.org/10.1371/JOURNAL.PONE.0312709

Glover JL, Champion D, Daniels KJ, Dainty AJD (2014) An Institutional Theory perspective on sustainable practices across the dairy supply chain. Int J Prod Econ 152(September):102–111. https://doi.org/10.1016/j.ijpe.2013.12.027

Graessler S, Guenter H, de Jong SB, Henning K (2024) Organizational change towards the circular economy: a systematic review of the literature. Int J Manag Rev 26(4):556–579. https://doi.org/10.1111/IJMR.12367

Hong T, Ou J, Jia F, Chen L, Yang Y (2024) Circular economy practices and corporate social responsibility performance: the role of sense-giving. Int J Log Res Appl 27(11):2208–2237. https://doi.org/10.1080/13675567.2023.2237914

Howe K, Lavy BL (2025) Business characteristics and leadership motivation in circular economy practices: a case study of small manufacturers in Dallas-Fort Worth. Front Sustain 6https://doi.org/10.3389/FRSUS.2025.1568099/FULL

Huertas López E, Suárez García E, Salgado Cruz M, Jadán Rodríguez LR, Jiménez Valero B (2020) Diseño de un modelo de gestión. Base científica y práctica para su elaboración. Universidad y Sociedad 2:1–9

ILO (2023) Decent work in the circular economy: an overview of the existing evidence base. https://www.ilo.org/publications/decent-work-circular-economy-overview-existing-evidence-base

Jensen HH (2025) Internal IT: the backbone system of circular business models. In: Circular economy opportunities and pathways for manufacturers. manufacturing renewed, pp 291–323. https://doi.org/10.1007/978-3-031-75279-7_12

Kantabutra S (2021) Exploring relationships among sustainability organizational culture components at a leading Asian industrial conglomerate. Sustainability 13(4):1–31. https://doi.org/10.3390/SU13041733

Karanfiloğlu M (2025) Digital literacy as a catalyst for sustainable development. In: Editordigital literacy as a catalyst for critical thinkingfrom media to artificial intelligence, pp 155–188. https://doi.org/10.1007/978-3-031-96720-7_8

Ketprapakorn N, Kantabutra S (2022) Toward an organizational theory of sustainability culture. Sustain Prod Consum 32:638–654. https://doi.org/10.1016/J.SPC.2022.05.020

Kopelman RK (2023) Understanding douglas McGregor's theory X and theory Y in human resource management. Glob Res J Soc Sci Manag 01(02):50–59. https://doi.org/10.55306/grjssm.2023.1205

Kurznack L, Schoenmaker D, Schramade W (2021) A model of long-term value creation. J Sustain Financ Invest 1–19https://doi.org/10.1080/20430795.2021.1920231

Llanos Encalada M (2016) La Personalidad De Las Organizaciones: La Cultura 7–24. https://libros.ecotec.edu.ec/index.php/editorial/catalog/download/53/48/714-1?inline=1

Loste N, Callao C, Giner B, Chinarro D (2017) El MOOC como instrumento para difundir el conocimiento sobre sostenibilidad y economía circular. CEUR Work Proc 1993:144–145

Lozano R (2015) A holistic perspective on corporate sustainability drivers. Corp Soc Responsib Environ Manag 22(1):32–44. https://doi.org/10.1002/CSR.1325;WGROUP:STRING:PUBLICATION

Ly B (2025) Beyond conventional HRM: catalyzing circular economy performance with green practices and inclusive leadership. Asia Pac Manag Rev.https://doi.org/10.1016/J.APMRV.2025.100373

Mahajan R, Lim WM, Sareen M, Kumar S, Panwar R (2023) Stakeholder theory. J Bus Res 166https://doi.org/10.1016/J.JBUSRES.2023.114104

Marrucci L, Daddi T, Iraldo F (2021) The contribution of green human resource management to the circular economy and performance of environmental certified organisations. J Clean Prod 319https://doi.org/10.1016/J.JCLEPRO.2021.128859

Mazzucchelli A, Chierici R, Del Giudice M, Bua I (2022) Do circular economy practices affect corporate performance? Evidence from Italian large-sized manufacturing firms. Corp Soc Responsib Environ Manag 29(6):2016–2029. https://doi.org/10.1002/CSR.2298

Mena Méndez D (2024) La cultura organizacional, elementos generales, mediaciones e impacto en el desarrollo integral de las instituciones. Revista científica Pensamiento y Gestión 46:11–47. https://doi.org/10.14482/pege.46.1203

Mogena Medina C, Prado-Solano A, Domínguez Durán J (2023) Economía circular en la formación para el empleo. Rev Digit Univ 24(5). https://doi.org/10.22201/CUAIEED.16076079E.2023.24.5.3

Morais DOC, Barbieri JC (2022) Supply chain social sustainability: unveiling focal firm's archetypes under the lens of stakeholder and contingency theory. Sustainability (Switzerland) 14(3):19–21. https://doi.org/10.3390/su14031185

Muhammad Rehman A, Ali Alqahtani D (2025) The moderating role of organizational culture on barriers and drivers of sustainable construction practices in Saudi Arabia's construction industry : a circular economy perspective. Buildings 15(3663):1–21. https://doi.org/10.3390/buildings15203663

Munonye WC (2025) Integrating circular economy principles in business strategies: a policy-driven approach. Circ Econ Sustain 5(3):1887–1896. https://doi.org/10.1007/S43615-025-00523-5/FIGURES/2

Muzondo PJ, Mashapure R, Masiiwa ST (2025) Barriers and enablers in adopting circular business models in horticulture SMEs: insights from Zimbabwe. J Circ Econ 3(3). https://doi.org/10.55845/GYQR3716

Ndaruhutse S, Jones C, Riggall A (2019) Why systems thinking is important for the education sector. Educ Dev Trust. http://search.ebscohost.com/login.aspx?direct=true&db=eric&AN=ED603263&site=ehost-live

Ogunmakinde OE, Memari A, Aghajani M (2025) Benchmarking international best practices for circular skills training: lessons for Australia. CIB Conf 1(1). https://doi.org/10.7771/3067-4883.1080

Olipp N, Schwarz M, Woschank M (2025) Integrating the principles of reverse logistics into circular economy strategies: a mixed-method study of small and medium-sized enterprises. Sustainability 17(16):7361. https://doi.org/10.3390/SU17167361

Opferkuch K, Caeiro S, Salomone R, Ramos TB (2021) Circular economy in corporate sustainability reporting: a review of organisational approaches. Bus Strat Environ 30(8). https://doi.org/10.1002/bse.2854

Park JH, Niu W, Cheng L, Allen H (2021) Fostering creativity and critical thinking in college: a cross-cultural investigation. Front Psychol 12.https://doi.org/10.3389/FPSYG.2021.760351/FULL

Pedrini S, Tartari M (2023) Cultural and creative practices as drivers of circular economy: a 9R framework. In: XI Computational science and its applications—ICCSA 2025 workshops, pp 283–299. https://doi.org/10.1007/978-3-031-97654-4

Piwowar-Sulej K, Iqbal Q (2023) Leadership styles and sustainable performance: a systematic literature review. J Clean Prod 382https://doi.org/10.1016/J.JCLEPRO.2022.134600

Renfors SM (2024) Education for the circular economy in higher education: an overview of the current state. Int J Sustain High Educ. https://doi.org/10.1108/IJSHE-07-2023-0270

Riemer M, Marcus J, Reimer-Watts BK (2025) Organizational journeys toward strong cultures of sustainability: a qualitative inquiry. Front Psychol 16(February):1–16. https://doi.org/10.3389/fpsyg.2025.1508818

Rizos V, Behrens A, van der Gaast W, Hofman E, Ioannou A, Kafyeke T et al (2016) Implementation of circular economy business models by small and medium-sized enterprises (SMEs): barriers and enablers. Sustainability (Switzerland) 8(11). https://doi.org/10.3390/su8111212

Rudolph CW, Rauvola RS, Zacher H (2018) Leadership and generations at work: a critical review. Leadersh Quart 29(1):44–57. https://doi.org/10.1016/J.LEAQUA.2017.09.004

Sadaoui N, Benlaria H, Almawishir NF, Abdelrhman BS, Ahmed SAK, Zabat L et al (2025) How technological innovation and organizational culture influence circular economy adoption over time in Saudi Arabian manufacturing. Discover Sustain 6(1). https://doi.org/10.1007/S43621-025-01662-3

Salvioni DM, Almici A (2020) Transitioning toward a circular economy: the impact of stakeholder engagement on sustainability culture. Sustainability 12(20):1–30. https://doi.org/10.3390/SU12208641

Sánchez Castro E (2021) Definición de cultura organizacional. Claiscos de la teoria organizacional 1–22. https://repositorio.usam.ac.cr/xmlui/handle/11506/1651?show=full

Santiago BS, Scavarda LF, Gusmão Caiado RG, Santos RS, Mattos Nascimento DL (2025) Corporate social responsibility and circular economy integration framework within sustainable supply chain management: building blocks for industry 5.0. Educ Adm: Theory Pract 32(1):269–290. https://doi.org/10.1002/CSR.2949

Singh P, Mishra GP (2024) Integrating corporate social responsibility (CSR) and circular economy (CE): a pathway to sustainable business practices and global resilience. Educ Adm: Theory Pract 30(11):1251–1259. https://doi.org/10.53555/kuey.v30i11.9303

Soto Kiewit LD, Segura Jiménez A, Navarro Rojas Ó, Cedeño Rojas S, Medina Díaz R (2023) Formal, non-formal and informal education and innovation: innovating to educate and educating to innovate. Rev Innovaciones Educ 25(38):77–96. https://doi.org/10.22458/IE.V25I38.4535

Srisathan WA, Ketkaew C, Naruetharadhol P (2020) The intervention of organizational sustainability in the effect of organizational culture on open innovation performance: a case of thai and chinese SMEs. Cogent Bus Manag 7(1). https://doi.org/10.1080/23311975.2020.1717408

Sulich A, Sołoducho-Pelc L, Ferasso M (2021) Management styles and decision-making: pro-ecological strategy approach. Sustainability 13(4):1–18. https://doi.org/10.3390/SU13041604

Teixeira N (2025) Circular economy perspectives: challenges, innovations, and sustainable futures. Discover Sustain 6(1). https://doi.org/10.1007/S43621-025-01606-X

Townsend M, Romme AGL (2024) The emerging concept of the human-centered organization: a review and synthesis of the literature. Humanistic Manage J 9(1):53–74. https://doi.org/10.1007/s41463-024-00168-w

Tudor-Tiron A, Dragu I (2014) From sustainability to integrated reporting—the political the political perspective of institutional theory. Studia Universitatis 59(2):59–78

UNESCO (2020) Educación para el Desarrollo Sostenible: Hoja de ruta. Organización de las Naciones Unidades para la Educación, la Ciencia y la Cultura

Wang H, Li J, Han P, Ouyang L (2023) The effects of differential organizational culture types on firm and individual performance in Eastern and Western cultural countries including China and the United States: a meta-analysis. Asia Pac Bus Rev 31(4):721–754. https://doi.org/10.1080/13602381.2023.2255560

Wasieleski D, Waddock S, Fort T, Guimarães-Costa N (2021) Natural sciences, management theory, and system transformation for sustainability. Bus Soc 60(1):7–25. https://doi.org/10.1177/0007650319898384

World Economic Forum (2025) Future of jobs report. Facilitating work-based learning.https://doi.org/10.1007/978-1-137-40325-4

Xu G, Wang S, Li J, Zhao D (2020) Moving towards sustainable purchase behavior: examining the determinants of consumers' intentions to adopt electric vehicles. Environ Sci Pollut Res 27(18):22535–22546. https://doi.org/10.1007/s11356-020-08835-9

Yücel Karamustafa E, Arsan B, Palalar Alkan D (2024) How can the circular economy contribute to the achievement of SDG 8: decent work and economic growth? In: Transition to the circular economy model. The case of Turkey, vol Part F2542. Springer Nature, pp 57–74. https://doi.org/10.1007/978-3-031-52700-5_4

Yusuf MF, Mohamad Nasarudin NAI, Sorooshian S, Fauzi MA, Kasim NM (2023) Exploring the impact of contingency theory on sustainable innovation in Malaysian manufacturing firms. Sustainability 15(9). https://doi.org/10.3390/su15097151

Chapter 6
Digitalization and Circular Economy in Industry 4.0

Abstract This chapter examines how digitalization and Industry 4.0 technologies are reshaping industrial systems by embedding circular economy principles into their core operations. It explores how digital tools—such as the Internet of Things, Artificial Intelligence, blockchain, and cloud computing—enable traceability, resource optimization, and lifecycle extension, thereby supporting the achievement of Sustainable Development Goal 12 (SDG 12) on responsible consumption and production. The chapter distinguishes between digitalization and digital transformation: while the former refers to the integration of digital technologies into existing processes, the latter implies systemic change across organizational structures and governance. In this context, digital tools are not merely operational enhancers but strategic enablers of circularity. Industry 4.0 technologies facilitate closed-loop supply chains, predictive maintenance, and intelligent waste valorization. Modular design and cyber-physical systems allow for product disassembly and reuse, while digital twins and lifecycle assessments support real-time monitoring and impact evaluation. Circular supply chain management integrates forward and reverse logistics, supported by big data analytics and collaborative platforms. The chapter also addresses critical challenges. The energy use of digital infrastructure—especially data centers and AI training—raises concerns about carbon emissions and water consumption. Material demands for digital devices contribute to electronic waste, often managed through informal and hazardous recycling methods. Digital divides persist across socioeconomic and geographic boundaries, limiting equitable access to technology and its sustainability benefits. Additionally, automation-driven job displacement calls for inclusive workforce strategies and ethical governance. Ultimately, the chapter argues that digitalization must be guided by principles of environmental responsibility, social justice, and inclusive policy frameworks to truly advance circular economy goals and redesign industrial systems toward sustainability.

Keywords Digitalization · Industry 4.0 · Blockchain · Artificial intelligence · Circular supply chains · Digital twins · Energy circularity · Digital divide

L. Velazquez, *Circular Economy and Sustainable Consumption in the Context of SDG 12*, Strategies for Sustainability, https://doi.org/10.1007/978-3-032-15598-6_6

6.1 Introduction

Digitalization and digital technologies are increasingly recognized as essential components of the evolving framework surrounding the circular economy. Many circular practices—such as resource tracking, product-as-a-service models, and reverse logistics—depend on digital infrastructure and data flow. According to the Organisation for Economic Co-operation and Development (OECD 2019), digitalization refers to "the use of digital technologies and data as well as interconnection that results in new or changes to existing activities," while digital transformation encompasses "the economic and societal effects of digitization and digitalization." These definitions clarify that digitalization enables operational changes, whereas transformation involves broader systemic shifts toward circularity. In this context, the World Bank Group (2024) in its Digital Progress and Trends Report highlights the rapid expansion of the digital sector: the Information Technology (IT) services segment has been the fastest growing part of the global economy, with a compound annual growth rate (CAGR) of 8% in value added and 6.7% in employment from 2000 to 2022—outpacing the global average of 5.1% and 1.2%, respectively. Additionally, global exports of IT services increased at 12% annually from 2010 to 2022, making it the most dynamic category in international trade in services.

In industrial contexts, this transformation is driven by a suite of technologies that enhance operational efficiency, innovation, and adaptability. The Internet of Things (IoT), Big Data analytics, and cloud computing enable real-time monitoring, data-driven decision-making, and scalable infrastructure, fostering process optimization and organizational resilience (Vărzaru and Bocean 2024). Artificial Intelligence (AI), robotics, and automation are reshaping industries by streamlining workflows, supporting predictive analytics, and prompting strategic workforce adaptation through new skill requirements and organizational restructuring (Babashahi et al. 2024). At the organizational level, success also depends on leadership, internal culture, and employee adaptability, with cloud-based tools and collaborative platforms facilitating agile communication and innovation across teams (Trenerry et al. 2021).

In the pursuit of sustainability, organizations are increasingly turning to Digital Transformation as a strategic enabler. Recent studies demonstrate how digital restructuring reshapes organizational processes and governance structures, fostering operational efficiency, transparency, and stakeholder engagement. Tang et al. (2025) show that digital tools contribute to carbon emission reductions and resource optimization, particularly in technology-intensive sectors. Zhang et al. (2022) emphasize its role in enhancing productivity, attracting talent, and promoting inclusive decision-making, Xia et al. (2022) identify key dimensions—such as innovation, supply chain modernization, and organizational agility—through which digital technologies support sustainability.

Adding to this view, Bindeeba et al. (2025) present a meta-analysis confirming a consistent positive relationship between DT and sustainable business outcomes across diverse sectors and regions. Their findings suggest that technologies like smart factories, big data analytics, and the Internet of Things are among the most influential enablers of sustainability, with particularly strong impacts in manufacturing and telecommunications. Rather than focusing solely on efficiency gains, organizations are now embracing systemic redesigns where digital innovation serves as the foundation for circular strategies. This approach enables a more holistic transformation, aligning technological capabilities with long-term sustainability objectives. As Ali et al. (2025) highlight, access to reliable digital data is essential for implementing circular economy strategies effectively, particularly in sectors with complex value chains. This foundational role of digitalization contributes to Sustainable Development Goal 12 by enabling resource efficiency and waste reduction, while also supporting SDG 8 through the promotion of decent work and inclusive economic growth.

Extending this perspective, Radhika et al. (2025) demonstrate that artificial intelligence-driven waste sorting and predictive maintenance significantly improve recycling efficiency and asset longevity, while blockchain technology ensures secure and transparent tracking of materials across supply chains, reducing counterfeiting and enhancing consumer trust. Their regression analysis reveals that blockchain adoption, AI integration, and circular supply chain activities have a statistically significant positive impact on recycling efficiency and waste reduction. Möller et al. (2025) complement this view by emphasizing that digital transformation technologies—such as modular design, digital tagging, and intelligent systems—play a foundational role in enabling circular economy strategies. These tools support the extension of product lifecycles, facilitate resource recovery, and enhance traceability across production and recycling processes, laying the groundwork for more sustainable and data-driven circular business models.

With digital tools now embedded in industrial ecosystems, regenerative resource flows, transparency, and adaptive governance are becoming central to circular economy strategies. This convergence is particularly evident in the rise of Industry 4.0 (I4.0), where technologies such as cyber-physical systems, smart factories, and interconnected platforms are redefining production and consumption models. These innovations lay the groundwork for data-driven circular business models, as highlighted by recent studies on modular design, intelligent systems, and traceability. The next section examines how I4.0 provides a strategic framework for operationalizing circular economy principles, aligning digital innovation with sustainable industrial transformation. While this chapter focuses on the enabling role of digitalization in advancing circularity, it also acknowledges the complex environmental, social, and ethical implications of these transformations—including energy demands, digital inequality, labor displacement, and governance challenges—which will be critically explored in subsequent sections.

6.2 Enabling Circularity Through Industry 4.0

The integration of advanced digital technologies within Industry 4.0 has become a cornerstone for reconfiguring industrial models and accelerating circularity. These technologies offer solutions to long-standing environmental and operational challenges, particularly in supply chain management. As Parra-Sánchez (2025) highlights, the adoption of IoT has evolved into a foundational platform for business transformation, with global investments surpassing $268 billion in 2022 and a projected annual growth rate of 15%. Han and Li (2025) demonstrate that blockchain, big data analytics, additive manufacturing, cloud computing, and cyber-physical systems collectively enhance transparency, resource optimization, and waste reduction across supply chains. These tools support closed-loop operations and responsive decision-making, while also addressing barriers such as data fragmentation, technological incompatibility, and eco-efficiency deficits. Furthermore, Sujatha et al. (2022) emphasize that smart production environments powered by machine learning and sensor networks can autonomously diagnose faults, optimize energy use, and adapt to shifting demands. Similarly, Briatore et al. (2025) underscore the role of Industry 4.0 in reducing carbon footprints and enabling lifecycle assessments, reinforcing its strategic relevance for sustainable industrial transformation.

Beyond technological efficiency, the transformative potential of Industry 4.0 lies in its capacity to integrate innovation with human-centered values and ecological responsibility. As Mayer and Oosthuizen (2022) emphasize, this digital revolution is not merely technological but deeply human-centered, requiring continuous skill development, contextual sensitivity, and ecological awareness. Their empirical study of a South African multinational engineering firm reveals that sustainable business practices in I4.0 contexts are most effective when technological advancement is balanced with human expertise, inclusive governance (e.g., Broad-Based Black Economic Empowerment), and environmental stewardship. In line with this, Jaskó and Ruppert (2025) highlight how I4.0 technologies support energy-efficient production, waste reduction, and the development of eco-friendly materials—contributing directly to net-zero carbon goals and aligning with SDG 12's emphasis on sustainable consumption and production.

For its part, Beier et al. (2021) critically assess the prevailing discourse around Industry 4.0 and its alignment with the Sustainable Development Goals, particularly SDG 12, on responsible consumption and production. Their systematic literature review reveals that while economic benefits such as productivity and growth are frequently cited, the environmental and social dimensions are often underrepresented or lack empirical substantiation. In response to these gaps, Sharma et al. (2024) present a conceptual framework that integrates Industry 4.0 technologies with circular economy practices (CEPs) to support the realization of SDG 12. Drawing on dynamic capabilities theory, institutional theory, and the triple bottom line perspective, the study emphasizes the synergistic convergence of digital transformation and sustainability-oriented strategies.

Industry 4.0 technologies—previously identified as key enablers—continue to support circular strategies through interconnected sensing, intelligent decision-making, and secure data flows (blockchain), each providing unique capabilities for industrial transformation. Additionally, the framework emphasizes how institutional pressures, including mimetic and normative forces, can shape firms' strategic direction toward adopting I4.0 and CEPs, thus integrating sustainability into core operational models. Ferraro et al. (2023) complement this approach by introducing a multi-criteria decision-making framework using the Analytic Hierarchy Process (AHP), a structured method that helps evaluate complex options by organizing decision factors into a hierarchy and assigning relative weights.

The transition to Industry 4.0 has enabled the digitalization of manufacturing processes, yielding improvements in quality control, operational agility, and data-driven decision-making. As these technologies become embedded within industrial ecosystems, their convergence with circular economy principles is reshaping production and consumption models. This integration fosters regenerative resource flows and enhances transparency—key elements for advancing SDG 12 (Rifqi et al. 2021). Smart services and intelligent production systems not only drive innovation but also reconfigure supply chains toward greater adaptability and sustainability (Atif et al. 2021; Mehmood et al. 2025), laying the groundwork for circular industrial transformation.

Circular economy principles—such as reuse, remanufacturing, and recycling—are increasingly supported by advanced digital technologies, including IoT sensors, cloud computing, and blockchain, which enable comprehensive tracking of materials and products throughout their lifecycle (Laskurain-Iturbe et al. 2021; Lei et al. 2023). Among these, blockchain stands out for its capacity to ensure secure, transparent, and traceable data flows across industrial supply chains, thereby facilitating the implementation of circular strategies and promoting sustainable business models. Additionally, it supports performance indicators such as Reduce Input Consumption (RIC) and Reduce Waste and Emission (RWE), which are central to achieving SDG 12. In regulatory contexts, the Eco-design for Sustainable Products Regulation (ESPR) mandates the use of Digital Product Passports (DPPs), reinforcing the role of digital tools in enhancing lifecycle transparency across complex value chains (Horn and Silvennoinen 2025).

Artificial intelligence is emerging as a transformative force in advancing circular economy strategies within Industry 4.0. Through intelligent data analysis, predictive modeling, and real-time decision-making, AI enables the optimization of resource use, waste management, and production efficiency. Abdallah et al. (2020) demonstrate its growing application in solid waste management, where computational techniques support the prediction of waste characteristics, container optimization, and logistics planning—contributing to material recovery and energy efficiency. Beyond operational improvements, AI-driven systems facilitate the design of sustainable products and processes by identifying opportunities for circular interventions. Integrated into smart factories and supply chains, AI enhances industrial adaptability and intelligence. These insights collectively reinforce the strategic alignment between digital innovation and SDG 12, complementing broader frameworks that connect circular economy principles and Industry 4.0 to new business models and the goals of the 2030 Agenda (Dantas et al. 2021).

6.3 Applying Digital Solutions for Circular Economy Transformation

Digital technologies are reshaping the foundations of circular economy implementation. While traceability systems, intelligent waste valorization, and energy circularity represent some of the most prominent applications, they are not the only ones. This section explores how these digital solutions enable measurable, scalable, and strategic transformations aligned with SDG 12.

6.3.1 Traceability and Lifecycle Monitoring

Digital traceability systems are increasingly recognized as strategic enablers of circular economy practices, especially within complex, multi-tier supply chains. These systems allow organizations to monitor material flows, verify recycled content, and ensure compliance with sustainability standards. Takkar et al. (2025) demonstrate that blockchain-based frameworks such as WEave can capture post-consumer lifecycle events—returns, recycling, and reuse—without compromising system performance. Their simulation validates the feasibility of monitoring recycled content and return processing rates through smart contract logic and traceability metrics. Complementary efforts across sectors reinforce this potential: Dos Santos Ferreira et al. (2024) propose a reverse logistics system for agricultural pesticide packaging that uses smart contracts and QR-coded tokens to incentivize proper recycling and ensure immutable lifecycle records. In construction, Isiaku and Dawud (2025) identify transparency and data integration as critical success factors for traceability systems in circular supply chains. Hakola et al. (2025) reinforce the role of Digital Product Passports in enabling traceability across electronic value chains. Their integration with smart tags—such as printed humidity sensors—allows real-time monitoring of environmental exposure, helping determine whether components are suitable for reuse or recycling.

While these innovations demonstrate significant potential, their strategic effectiveness depends on deeper integration with sustainability frameworks and governance models. Santana and Ribeiro (2022) caution that most traceability models and systems remain disconnected from environmental, social, and economic indicators. Their systematic review of 49 studies reveals that current implementations often lack ontological frameworks and robust monitoring tools, which are essential for consistency, transparency, and accountability across sectors. They argue that traceability must evolve beyond technical tracking to become a strategic instrument of circular governance—one that integrates lifecycle data with sustainability metrics and supports cross-sectoral decision-making. Without such integration, traceability risks becoming a fragmented tool, insufficient to meet the demands of circular transformation. Celik et al. (2024) reinforce this concern by highlighting that traceability systems in construction, while technically robust, often lack integration with standardized sustainability frameworks. Their review identifies fragmentation, limited interoperability, and the absence of sector-specific platforms as key barriers to using traceability as a strategic governance tool. These limitations highlight the importance of complementing traceability systems with intelligent material management strategies—an area where digital tools continue to evolve.

6.3.2 *Intelligent Material Separation and Waste Valorization*

The shift from traditional waste management to intelligent, digitally powered valorization systems is transforming how materials are sorted, recovered, and reintroduced into circular flows. Digital twins improve resource recovery through real-time monitoring and predictive analytics, lowering waste and energy use while increasing lifecycle transparency (Gupta 2025). Campana et al. (2025) emphasize the integration of AI and digital twins in waste forecasting and environmental monitoring, highlighting thematic clusters that drive innovation. Alexakis et al. (2025) present a breakthrough in high-precision classification by combining hyperspectral and RGB imaging, enabling accurate sorting even with industrial contamination. Ibitoye et al. (2025) propose a unified architecture combining IoT, AI, and blockchain across sensing, processing, and trust layers. This system allows traceability, automated compliance, and token-based rewards—digital incentives recorded on blockchain to motivate citizen participation in proper waste sorting. Hanes et al. (2021) propose the potential of data-driven circular interventions through the Circular Economy Lifecycle Assessment for Valorization and Integration (CELAVI) framework, showing how mechanical recycling and cement co-processing retained up to 70% of wind turbine blade material in circulation, reducing landfill reliance and extending lifecycles despite a slight increase in emissions.

Bułkowska et al. (2023) support these advances by showing how blockchain-based smart contracts and digital product passports can automate verification, trace waste origins, and optimize logistics in real time. Their Waste 24 case study illustrates how IoT sensors linked with blockchain reduce operational costs and improve transparency in municipal waste collection, providing a scalable model for smart cities. As Calderon-Monge and Ribeiro-Soriano (2024) note, digital tools facilitate the recovery and exchange of non-biodegradable waste via mobile apps, promote eco-efficient manufacturing through robotics and 3D printing, and enhance accountability with blockchain-enabled reporting systems. However, despite their transformative potential, these systems face persistent structural, technological, and socio-cultural barriers. Frantz Schneider et al. (2024), in a multi-case study of formal e-waste organizations in Brazil, identify obstacles such as limited infrastructure, high reverse logistics costs, and weak policy integration. Kathi et al. (2023) highlight technological bottlenecks in wastewater valorization, including membrane fouling, enzyme instability, and scalability issues in microalgae-based systems, alongside consumer skepticism and lack of fiscal incentives. Ayeni (2025) and Fuqaha and Nursetiawan (2025) further emphasize the need for coordinated governance, ethical frameworks, and inclusive policy design to overcome fragmented infrastructure, algorithmic bias, and social inequity.

6.3.3 Energy Circularity and Carbon Reduction

Energy circularity serves as a strategic enabler for achieving SDG 12 by fostering resource efficiency, lifecycle extension, and carbon mitigation across energy systems. Oladapo et al. (2024) provide empirical evidence that circular practices in industrial energy operations—such as refining and electric vehicle manufacturing—can reduce waste by 25% and boost recycling efficiency from 50 to 83%, supported by predictive models like BLUES and ARIMA with less than 5% deviation from actual performance. Munonye (2025) explores how integrating circular economy principles with smart energy grids enhances resilience and sustainability through decentralized energy systems, prosumer engagement, and localized reuse of energy resources. Al-Shetwi et al. (2025) demonstrate the transformative impact of digital twin technology in optimizing renewable energy integration, energy storage, and vehicle-to-grid systems, enabling real-time monitoring and predictive maintenance that improve grid stability and reduce emissions—aligning digital innovation with circular energy transitions.

Building on the strategic role of energy circularity in SDG 12, waste-to-energy (WtE) technologies offer a complementary pathway by converting residual waste into usable energy, thereby reducing landfill dependency and enhancing resource recovery. Rezania et al. (2023) emphasize that WtE systems—particularly incineration, anaerobic digestion, and gasification—are increasingly integrated with digital infrastructure such as waste flow tracking and environmental impact assessments, enabling more precise control over emissions and energy outputs in both developed and developing countries. Themelis (2023) provides a longitudinal analysis of WtE adoption, showing that nations like Denmark and Japan have successfully phased out landfilling by deploying advanced combustion systems equipped with automated air pollution control technologies, including activated carbon injection and baghouse filters, which reduce dioxin emissions by a factor of 1000. Inayat et al. (2025) explore hybrid WtE systems that combine anaerobic digestion and gasification, highlighting how digital reactor modeling and sensor-based monitoring optimize feedstock variability and energy yield, achieving up to 90.2% system efficiency and a 161% increase in biomethane production when integrated with membrane separation and catalytic conversion technologies. Abdul-Hamid et al. (2021) reinforce this by showing that energy recovery from palm-based residues—used in biorefineries and eco-industrial parks—not only reduces greenhouse gas (GHG) emissions but also enhances economic viability by supplying electricity to processing facilities and lowering operational costs.

6.3.4 Modular Design and Product Lifecycle Extension

Modular design and cyber-physical systems are key enablers of circularity, allowing products to be disassembled into reusable components and integrated into digital monitoring frameworks. Frank et al. (2019) highlight how these technologies—central to Industry 4.0—support smart manufacturing and lifecycle transparency, which are essential for implementing circular strategies in industrial contexts. In the aerospace sector, turbine blades have been successfully remanufactured using Laser Direct Deposition (LDD), which allows for precise reconstruction of damaged regions and results in up to 45% carbon footprint reduction compared to full replacement. Similarly, in the automotive industry, remanufactured diesel engines have demonstrated a 66% reduction in energy consumption and up to 97% lower ozone depletion potential. In port logistics, rotary-type handling equipment has been retrofitted with Actual State Monitoring (ASM) systems, improving maintenance planning and reducing operational costs by up to 60%. These applications showcase the environmental and economic benefits of life extension strategies and are supported by digital tools such as Lifecycle Assessment (LCA) and predictive maintenance systems, which help quantify impacts and optimize interventions (Ferreira and Gonçalves 2021).

Beyond traditional sectors, digitalization has enabled new models for product life extension. For example, the use of digital twins and sensors from the Internet of Things in industrial equipment allows for real-time monitoring of wear and performance, facilitating timely refurbishment and reducing downtime (Fontana et al. 2021). Moreover, servitization models such as Product-as-a-Service (PaaS) are gaining traction, where manufacturers retain ownership and responsibility for maintenance, enabling continuous upgrades and reuse cycles (Kazakova and Lee 2022). Organizations that adopt technologies such as Artificial Intelligence and blockchain report up to a 25% improvement in operational sustainability, including reductions of 30% in energy consumption and 20% in material waste, demonstrating the measurable impact of digital tools in circular economy strategies (Setyadi et al. 2025).

6.3.5 Circular Supply Chain Management

Circular supply chain management (CSCM) reflects a paradigm shift from linear logistics models toward regenerative systems that prioritize resource efficiency and waste minimization. Lopes de Sousa Jabbour et al. (2018) emphasize that Industry 4.0 technologies enable modular product design and closed-loop manufacturing, which are essential for circularity. Furthermore, Choi and Chen (2021) propose that big data analytics and large-scale group decision-making frameworks are critical for coordinating diverse stakeholders and optimizing circular operations. This integration of digital tools with circular strategies enhances transparency and responsiveness across supply chains. As Bimpizas-Pinis et al. (2022) demonstrate, the concept of "arcs of integration" enables the simultaneous evaluation of forward and reverse flows, which is vital for managing complex networks involving recyclers, remanufacturers, and waste handlers.

In practice, circular supply chain management requires not only technological enablers but also strategic alignment across organizational functions. Cano et al. (2025) propose a comprehensive framework that integrates business models, collaboration, design, and technology as key enablers of circularity, linking them to measurable outcomes such as material, water, and energy flows. Their model operationalizes circular strategies through the 10Rs, offering a structured approach to assess circular performance across supply chain stages. Romero et al. (2021) reinforce this perspective by highlighting how Industry 4.0 technologies extend product lifespans and optimize resource use, particularly when applied to reverse logistics and sustainable sourcing. These insights underscore the importance of supply chain integration (SCI) not only as a technical mechanism but also as a cultural and strategic driver of transformation.

While digital tools offer unprecedented capabilities for circular transformation, their success depends on integration with governance frameworks, inclusive policies, and cross-sectoral collaboration. Achieving SDG 12 requires not only innovation, but systemic alignment and cultural commitment.

6.4 Critical Reflections on Digitalization for Circularity

While digitalization promises cleaner, more efficient industrial systems and transformative potential for circular economy strategies, it also introduces complex challenges that must be critically examined. The infrastructure that supports digital transformation—such as data centers, global platforms, and intelligent systems—demands vast energy resources and contributes to environmental pressures. At the same time, the rapid turnover of digital devices has intensified the problem of electronic waste, raising concerns about toxic exposure, informal recycling, and the ethical implications of global disposal patterns. These unintended consequences are compounded by persistent digital divides, which limit equitable access to technological benefits and deepen socioeconomic disparities. Moreover, the automation of tasks and the rise of intelligent systems are reshaping labor markets, prompting urgent debates around workforce reskilling and inclusion. Ethical concerns related to data governance, algorithmic bias, and surveillance further complicate the digital landscape. To ensure that digitalization genuinely supports circular economy goals, it must be guided by principles of environmental responsibility, social justice, and inclusive governance.

6.4.1 Energy Impact of Digital Infrastructure

The environmental cost of digitalization is increasingly concentrated in its energy-intensive infrastructure. According to the Digital Economy Report by United Nations Conference on Trade and Development (UNCTAD 2024), data centers consumed approximately 460 terawatt-hours (TWh) of electricity in 2022—a figure projected to double by 2026, potentially surpassing the total electricity consumption of countries like France. This surge is driven by the proliferation of compute-intensive technologies such as artificial intelligence, blockchain, and 5G networks, which demand vast amounts of processing power and storage. For instance, training large AI models like GPT-3 has been estimated to consume 700,000 L of potable water for cooling in a single data center. Moreover, cryptocurrency mining, particularly under proof-of-work mechanisms, reached 121 TWh in energy consumption by 2023, a 34-fold increase since 2015. These trends not only strain global energy systems but also raise concerns about carbon emissions, water usage, and local environmental degradation, especially in regions where data centers are concentrated. According to the International Energy Agency (IEA 2025), global electricity consumption from data centers is expected to be more than double again, reaching 945 TWh by 2030 and potentially 1200 TWh by 2035, with AI-focused facilities consuming as much electricity as 100,000 households. Complementarily, the Bipartisan Policy Center and Koomey Analytics (2025) report notes that while national electricity demand in the United States has remained relatively stable, certain regions are experiencing substantial surges driven by concentrated data center development, highlighting the uneven and localized nature of digital infrastructure's energy impact.

6.4.2 Material Intensity and E-Waste Generation

As previously noted in the Digital Economy Report 2024, digitalization is heavily reliant on material resources. Producing a single 2 kg computer requires 800 kg of raw materials, including critical minerals such as lithium, cobalt, and rare earth elements—many of which come from regions with fragile ecosystems and vulnerable labor conditions. The demand for these materials is expected to increase by up to 500% by 2050, fueled by both digital and low-carbon technologies. This shift marks a transition from fossil fuel dependence to a multi-element reliance, with smartphones now containing up to 63 elements from the periodic table, many in trace amounts but requiring high purity levels that entail energy-intensive processing according to the United Nations Environment Programme (UNEP 2024). The environmental footprint of digitalization is further amplified by the rapid expansion of AI compute and data infrastructure, which intensifies energy and material demands. According to the OECD (2022), AI workloads are growing faster than hardware efficiency gains, raising concerns about long-term sustainability. This heightens the environmental impact of digitalization, especially during the manufacturing stage, which makes up to 80% of a device's lifecycle emissions.

When these devices reach end-of-life, they add to the growing e-waste problem, often ending up in landfills or informal recycling processes that lack proper environmental safeguards. According to the United Nations University (UNU), the United Nations Environment Programme, and the Solving the E-waste Problem Initiative (StEP), around 50 million metric tons of e-waste are generated globally each year, with projections reaching 75 million tons by 2030 and 111 million tons by 2050 (Parajuly et al. 2019). Much of this waste is exported from industrialized nations to developing countries, where informal recyclers use rudimentary techniques that not only fail to recover valuable materials but also cause severe environmental and health hazards due to toxic emissions and improper handling. The International Telecommunication Union (ITU) highlights that the rapid expansion of digital infrastructure—especially data centers, IoT devices, and 5G networks—is accelerating the volume and complexity of e-waste, with many components containing critical raw materials that are difficult to recover and often lost in landfills or incineration (ITU 2020).

6.4.3 Digital Divides

Digital inequality among individuals and communities remains a critical barrier to inclusive development and sustainability. Access to infrastructure, devices, and digital skills is unevenly distributed across socioeconomic, geographic, generational, and gender lines. According to the World Bank Group (2023), women, rural populations, and persons with disabilities face compounded barriers that limit their participation in the digital economy and access to essential services. In India, Laskar (2023) identifies how under-resourced urban and rural areas lack the infrastructure and literacy needed to benefit from digitalization, reinforcing existing social inequalities. McCall et al. (2022) highlight that in the United States, digital redlining has deepened disparities in education and healthcare, particularly during the COVID-19 pandemic. These divides not only restrict access to information and services but also perpetuate broader patterns of exclusion.

This exclusion is mirrored in the industrial sector, where uneven digital adoption creates structural divides that hinder progress toward Sustainable Development Goal 12. Large corporations have integrated tools such as artificial intelligence, cloud computing, and big data analytics, while small- and medium-sized enterprises (SMEs) often face limited infrastructure, low digital skills, and restricted access to financing. According to the International Labour Organization (ILO 2021), only 20% of SMEs in low-income countries use websites or digital platforms, despite representing over 90% of employment in some regions. Travassos Rosário and Carmo Dias (2023) warn that digitalization risks concentrating sustainability benefits in a few sectors, excluding traditional industries such as agriculture and manufacturing. Misra and Srivastava (2024) emphasize that these sectors lack the capabilities needed to implement circular models and reduce environmental impact. Bridging these gaps is essential to ensure that all industries and communities contribute meaningfully to sustainability—upholding the 2030 Agenda's commitment to "leave no one behind."

6.4.4 Labor Displacement and Protection in the Era of Industry 4.0

The expansion of digital technologies, particularly artificial intelligence, has intensified debates around labor displacement and the future of work. While automation is often associated with large-scale job loss, recent evidence suggests a more nuanced reality. Willcocks (2024) argues that automation is increasingly deployed as a coping mechanism to manage the exponential growth in work volume, driven by data proliferation, regulatory complexity, and technological interdependence. The real challenge lies in the dramatic shift in skills demand: low-skilled roles are declining, while the demand for digital, cognitive, and uniquely human skills—such as empathy, critical thinking, and leadership—is rising. Moncada et al. (2025) reinforce this trend, showing that digital investments in manufacturing sectors correlate positively with the hiring of workers holding post-secondary technical diplomas (ITS) and postgraduate STEM qualifications, while negatively affecting the recruitment of individuals with only secondary education. However, access to these emerging opportunities is not equitably distributed. McCall et al. (2022) highlight that broadband Internet access—now recognized as a social determinant of health—is systematically denied to low-income and racialized communities through digital redlining, thereby limiting their access to education, employment, and healthcare. To mitigate these risks, Mukhuty et al. (2022) propose socially responsible human resource management (HRM) strategies, including inclusive talent development, collaboration with educational institutions to redesign curricula, and the use of smart technologies to identify skill gaps and support internal redeployment. Without proactive governance and inclusive digital infrastructure, digitalization risks deepening inequalities and undermining decent work.

6.5 Conclusion

This chapter has explored how digitalization and Industry 4.0 technologies serve as strategic enablers of circular economy transformation, particularly in the context of Sustainable Development Goal 12. By integrating tools such as artificial intelligence, blockchain, digital twins, and modular design, industrial systems are increasingly capable of achieving resource efficiency, lifecycle extension, and waste reduction. The convergence of digital innovation and circular principles is reshaping supply chains, product design, and energy systems, offering scalable solutions for sustainable production and consumption.

However, the chapter also underscores the critical need to address the unintended consequences of digitalization, including energy-intensive infrastructure, electronic waste, digital inequality, and labor displacement. These challenges highlight the importance of embedding ethical governance, inclusive policy frameworks, and environmental responsibility into digital strategies.

Ultimately, the chapter argues that digitalization must move beyond operational enhancement to become a vehicle for systemic transformation. When guided by sustainability values and supported by cross-sectoral collaboration, digital tools can help redesign industrial models toward regenerative, equitable, and resilient futures. This perspective reinforces the broader thesis of the book: that circular economy implementation requires not only technological innovation but also cultural, institutional, and strategic alignment across all levels of society.

References

Abdallah M, Abu Talib M, Feroz S, Nasir Q, Abdalla H, Mahfood B (2020) Artificial intelligence applications in solid waste management: a systematic research review. Waste Managehttps://doi.org/10.1016/j.wasman.2020.04.057

Abdul-Hamid AQ, Ali MH, Osman LH, Tseng ML (2021) The drivers of industry 4.0 in a circular economy: the palm oil industry in Malaysia. J Clean Prod 324:129216. https://doi.org/10.1016/J.JCLEPRO.2021.129216

Al-Shetwi AQ, Atawi IE, El-Hameed MA, Abuelrub A (2025) Digital twin technology for renewable energy, smart grids, energy storage and vehicle-to-grid integration: advancements, applications, key players, challenges and future perspectives in modernising sustainable grids. IET Smart Grid 8(1). https://doi.org/10.1049/STG2.70026

Alexakis G, Pellegrino M, Rodriguez-Turienzo L, Maniadakis M (2025) Enhanced waste sorting technology by integrating hyperspectral and RGB imaging. Recycling 10(5):179. https://doi.org/10.3390/RECYCLING10050179

Ali ZA, Zain M, Hasan R, Pathan MS, AlSalman H, Almisned FA (2025) Digital twins: cornerstone to circular economy and sustainability goals. Environ Dev Sustain.https://doi.org/10.1007/S10668-025-06221-4

Atif S, Ahmed S, Wasim M, Zeb B, Pervez Z, Quinn L (2021) Towards a conceptual development of industry 4.0, servitisation, and circular economy: a systematic literature review. Sustainability (Switzerland) 13(11). https://doi.org/10.3390/su13116501

Ayeni O (2025) Smart waste management in the age of industry 4.0: IoT, AI and blockchain applications in circular economy systems. World J Adv Res Rev 28(1):379–391. https://doi.org/10.30574/WJARR.2025.28.1.3358

Babashahi L, Barbosa CE, Lima Y, Lyra A, Salazar H, Argôlo M et al (2024) AI in the workplace: a systematic review of skill transformation in the industry. Adm Sc 14(6). https://doi.org/10.3390/ADMSCI14060127

Beier G, Niehoff S, Hoffmann M (2021) Industry 4.0: a step towards achieving the SDGs? A critical literature review. Discover Sustain 2(1). https://doi.org/10.1007/S43621-021-00030-1

Bimpizas-Pinis M, Calzolari T, Genovese A (2022) Exploring the transition towards circular supply chains through the arcs of integration. Int J Prod Econ 250.https://doi.org/10.1016/J.IJPE.2022.108666

Bindeeba DS, Tukamushaba EK, Bakashaba R (2025) Digital transformation and its multidimensional impact on sustainable business performance: evidence from a meta-analytic review. Futur Bus J 11(1). https://doi.org/10.1186/S43093-025-00511-Z

Bipartisan Policy Center, & Koomey Analytics (2025) Electricity demand growth and data centers: a guide for the perplexed. https://bipartisanpolicy.org/report/electricity-demand-growth-and-data-centers/

Briatore F, Vanni F, Mosca MT, Mosca RN, Fruggiero F, Mancusi F (2025) Exploring industry 4.0's role in sustainable supply chains: perspectives from a bibliometric review. Logistics 9(1). https://doi.org/10.3390/LOGISTICS9010026

Bułkowska K, Zielińska M, Bułkowski M (2023) Implementation of blockchain technology in waste management. Energies 16(23). https://doi.org/10.3390/EN16237742

Calderon-Monge E, Ribeiro-Soriano D (2024) The role of digitalization in business and management: a systematic literature review. RMS 18(2):449–491. https://doi.org/10.1007/S11846-023-00647-8

Campana P, Censi R, Tarola AM, Ruggieri R (2025) Artificial intelligence and digital twins for sustainable waste management: a bibliometric and thematic review. Appl Sci 15(11). https://doi.org/10.3390/APP15116337

Cano JA, Londoño-Pineda A, Campo EA, Gruchmann T, Weyers S (2025) Circular supply chain management assessment: a systematic literature review. Environments 12(10):374. https://doi.org/10.3390/ENVIRONMENTS12100374

Celik BG, Abraham YS, Attaran M (2024) Unlocking blockchain in construction: a systematic review of applications and barriers. Buildings 14(6). https://doi.org/10.3390/BUILDINGS14061600

Choi TM, Chen Y (2021) Circular supply chain management with large scale group decision making in the big data era: the macro-micro model. Technol Forecast Soc Chang 169https://doi.org/10.1016/J.TECHFORE.2021.120791

Dantas TET, de-Souza ED, Destro IR, Hammes G, Rodriguez CMT, Soares SR (2021) How the combination of circular economy and industry 4.0 can contribute towards achieving the sustainable development goals. Sustain Prod Consum 26:213–227. https://doi.org/10.1016/J.SPC.2020.10.005

Dos Santos Ferreira C, Gomes G, Castillo-Ospina DA, Bertassini AC, Zaguetto C, Feltrin N et al (2024) Organizational practices, values, and mindsets as a basis for circular economy transition. In: A systemic transition to circular economy, pp 85–110. https://doi.org/10.1007/978-3-031-55036-2

Ferraro S, Cantini A, Leoni L, De Carlo F (2023) Sustainable logistics 4.0: a study on selecting the best technology for internal material handling. Sustainability 15(9). https://doi.org/10.3390/SU15097067

Ferreira C, Gonçalves G (2021) A systematic review on life extension strategies in industry: the case of remanufacturing and refurbishment. Electronics 10(21). https://doi.org/10.3390/ELECTRONICS10212669

Fontana A, Barni A, Leone D, Spirito M, Tringale A, Ferraris M et al (2021) Circular economy strategies for equipment lifetime extension: a systematic review. Sustainability 13(3):1–27. https://doi.org/10.3390/SU13031117

Frank AG, Dalenogare LS, Ayala NF (2019) Industry 4.0 technologies: implementation patterns in manufacturing companies. Int J Prod Econ 210:15–26. https://doi.org/10.1016/j.ijpe.2019.01.004

Frantz Schneider A, Aanestad M, Carvalho TC (2024) Exploring barriers in the transition toward an established e-waste management system in Brazil: a multiple-case study of the formal sector. Environ, Dev Sustain.https://doi.org/10.1007/S10668-024-05188-Y

Fuqaha S, Nursetiawan N (2025) Artificial intelligence and IoT for smart waste management: challenges, opportunities, and future directions. J Futur Artif Intell Technol 2(1):24–46. https://doi.org/10.62411/FAITH.3048-3719-85

Gupta S (2025) Digital twins for circular economy optimization: a framework for sustainable engineering systems. Proceeding Pap 121(4). https://doi.org/10.3390/proceedings2025121004
Hakola L, Abedi F, Nordman S, Smolander M, Paltakari J (2025) Smart tags as enablers for digital product passports in circular electronics value chains. Circ Econ Sustain 5(3):2033–2055. https://doi.org/10.1007/S43615-025-00500-Y
Han H, Li J (2025) Leveraging industry 4.0 technologies for sustainable supply chain management: a systematic review. Clean Logist Supply Chain 16. https://doi.org/10.1016/J.CLSCN.2025.100246
Hanes R, Ghosh T, Key A, Eberle A (2021) The circular economy lifecycle assessment and visualization framework: a case study of wind blade circularity in Texas. Front Sustain 2https://doi.org/10.3389/FRSUS.2021.671979/FULL
Horn S, Silvennoinen K (2025) Traceability as an enabler of circular economy in the plastics packaging value chain. Circ Econ Sustain.https://doi.org/10.1007/S43615-025-00626-Z
Ibitoye SE, Ball AK, Rodriguez RV, Olayemi OA, Esteva P, Omoniyi PO et al (2025) Smart waste, smarter world: exploring waste types, trends, and tech-driven valorization through artificial intelligence, internet of things, and blockchain. Sustain Dev. https://doi.org/10.1002/SD.70345
IEA (2025) World energy outlook special report. Energy and AI. https://www.iea.org/reports/energy-and-ai/
ILO (2021) Small goes digital : how digitalization can bring about productive growth for micro and small enterprises. ILO. https://www.ilo.org/publications/small-goes-digital. Accessed 25 Oct 2025
Inayat A, Dafalla M, Asaad S, Jamil F, Al-Haj L, Shah FM et al (2025) Sustainable energy production from waste: a review of hybrid approaches combining anaerobic digestion and gasification. Int J Energy Res 2025(1). https://doi.org/10.1155/ER/6644084
Isiaku L, Dawud AM (2025) Blockchain and the smart circular economy: real-world applications and insight. In: Blockchain technologies for smart circular economy and organisational sustainability, pp 195–220. Springer Nature. https://doi.org/10.1007/978-3-031-89183-0_13
ITU (2020) Internet Waste. http://www.wipo.int/amc/en/mediation/rules
Jaskó S, Ruppert T (2025) The future of manufacturing and industry 4.0. Appl Sci 15(9). https://doi.org/10.3390/APP15094655
Kathi S, Singh S, Yadav R, Singh AN, Mahmoud AED (2023) Wastewater and sludge valorisation: a novel approach for treatment and resource recovery to achieve circular economy concept. Front Chem Eng 5https://doi.org/10.3389/FCENG.2023.1129783
Kazakova E, Lee J (2022) Sustainable manufacturing for a circular economy. Sustainability (Switzerland) 14(24). https://doi.org/10.3390/SU142417010
Laskar MH (2023) Examining the emergence of digital society and the digital divide in India: a comparative evaluation between urban and rural areas. Front Sociol 8.https://doi.org/10.3389/FSOC.2023.1145221/FULL
Laskurain-Iturbe I, Arana-Landín G, Landeta-Manzano B, Uriarte-Gallastegi N (2021) Exploring the influence of industry 4.0 technologies on the circular economy. J Clean Prod 321(September). https://doi.org/10.1016/j.jclepro.2021.128944
Lei Z, Cai S, Cui L, Wu L, Liu Y (2023) How do different Industry 4.0 technologies support certain circular economy practices? Ind Manag Data Syst 123(4). https://doi.org/10.1108/IMDS-05-2022-0270
Lopes de Sousa Jabbour AB, Jabbour CJC, Godinho Filho M, Roubaud D (2018) Industry 4.0 and the circular economy: a proposed research agenda and original roadmap for sustainable operations. Ann Oper Res 270(1–2):273–286. https://doi.org/10.1007/S10479-018-2772-8/METRICS
Mayer CH, Oosthuizen RM (2022) Sustainability in industry 4.0 business practice: insights from a multinational technology company. Front Sustain 3. https://doi.org/10.3389/FRSUS.2022.886986/FULL
McCall T, Asuzu K, Oladele CR, Leung TI, Wang KH (2022) A socio-ecological approach to addressing digital redlining in the United States: a call to action for health equity. Front Digital Health 4https://doi.org/10.3389/FDGTH.2022.897250/FULL

Mehmood K, Kautish P, Rashid M, Joshi Y, Iftikhar Y (2025) Digitalization in the circular economy: synergistic impact of big data analytics, green internet of things, and ambidextrous green innovation. J Clean Prod 509:145610. https://doi.org/10.1016/J.JCLEPRO.2025.145610

Misra P, Srivastava R (2024) Digital divide and sustainable development. In: Digital technologies to implement the UN sustainable development goals. Springer Nature. https://doi.org/10.1007/978-3-031-68427-2_23

Möller DPF, Wang Q, Huang L, Xiong B, Guo Y, Shi T, Zhang R (2025) Impact of digital transformation for sustainable circular economic environments. Carbon Neutral Systems 1(1). https://doi.org/10.1007/S44438-025-00001-3

Moncada R, Carbonero F, Geuna A, Riso L (2025) Digital adoption and human capital upscaling: a regional study of the manufacturing sector. Small Bus Econ 64(4):2061–2103. https://doi.org/10.1007/S11187-024-00975-3

Mukhuty S, Upadhyay A, Rothwell H (2022) Strategic sustainable development of industry 4.0 through the lens of social responsibility: the role of human resource practices. Bus Strat Environ 31(5):2068–2081. https://doi.org/10.1002/BSE.3008

Munonye WC (2025) Integrating circular economy principles in business strategies: a policy-driven approach. Circ Econ Sustain 5(3):1887–1896. https://doi.org/10.1007/S43615-025-00523-5/FIGURES/2

OECD (2019) Going digital: shaping policies, improving lives. going digital: shaping policies, improving lives. OECD. https://doi.org/10.1787/9789264312012-EN

OECD (2022) Measuring the environmental impacts of artificial intelligence compute and applications: the AI footprint. OECD Digit Econ Pap (341):1–11, 15–56. http://www.oecd.org/termsandconditions.%0Ahttps://ezproxy.cc.lut.fi/reports/measuring-environmental-impacts-artificial/docview/2737507074/se-2?accountid=27292%0Ahttps://libkey.io/libraries/2418/openurl?genre=unknown&au=Anonymous&aulast=Anonymous&issn=&isb

Oladapo BI, Olawumi MA, Omigbodun FT (2024) AI-driven circular economy of enhancing sustainability and efficiency in industrial operations. Sustainability 16(23). https://doi.org/10.3390/SU162310358

Parajuly K, Kuehr R, Awasthi AK, Fitzpatrick C, Lepawsky J, Smith E et al (2019) Future E-waste scenarios. UNU and UNEP-IETC. https://www.unep.org/ietc/resources/publication/future-e-waste-scenarios#:~:text=Thispapermapsseveraldimensionsofthee-waste,ofe-productsandmanagementofe-wastecouldevolve

Parra-Sánchez DT (2025) Exploring the internet of things adoption in the fourth industrial revolution: a comprehensive scientometric analysis. J Innov Digit Transform 2(1):1–18. https://doi.org/10.1108/JIDT-06-2024-0013

Radhika K, Uthaman VS, Murgai A, Dnyaneshwar VD, Kumar Singh A, Borah S (2025) Driving circular economy strategies through digital transformation and emerging technologies. J Inf Syst Eng Manag 10(30s):891–901. https://doi.org/10.52783/jisem.v10i30s.4939

Rezania S, Oryani B, Nasrollahi VR, Darajeh N, Lotfi Ghahroud M, Mehranzamir K (2023) Review on waste-to-energy approaches toward a circular economy in developed and developing countries. Processes 11(9). https://doi.org/10.3390/PR11092566

Rifqi H, Zamma A, Souda SB, Hansali M (2021) Positive effect of industry 4.0 on quality and operations management. Int J Online Biomed Eng 17(9):133–147. https://doi.org/10.3991/ijoe.v17i09.24717

Romero CAT, Castro DF, Ortiz JH, Khalaf OI, Vargas MA (2021) Synergy between circular economy and industry 4.0: a literature review. Sustainability 13(8):1–18. https://doi.org/10.3390/su13084331

Santana S, Ribeiro A (2022) Traceability models and traceability systems to accelerate the transition to a circular economy: a systematic review. Sustainability 14(9). https://doi.org/10.3390/SU14095469

Setyadi A, Soekotjo S, Lestari SD, Pawirosumarto S, Damaris A (2025) Trends and opportunities in sustainable manufacturing: a systematic review of key dimensions from 2019 to 2024. Sustainability 17(2). https://doi.org/10.3390/SU17020789

Sharma M, Singh P, Tsagarakis K (2024) Strategic pathways to achieve sustainable development goal 12 through industry 4.0: moderating role of institutional pressure. Bus Strat Environ 33(6):5812–5838. https://doi.org/10.1002/BSE.3769

Sujatha M, Priya N, Beno A, Blesslin Sheeba T, Manikandan M, Tresa IM et al (2022) IoT and Machine learning-based smart automation system for industry 4.0 using robotics and sensors. J Nanomater 2022. https://doi.org/10.1155/2022/6807585

Takkar R, Birman K, Gao HO (2025) Enhancing transparency in buyer-driven commodity chains for complex products: extending a blockchain-based traceability framework towards the circular economy. Appl Sci 15(15). https://doi.org/10.3390/APP15158226

Tang J, Huang K, Xiong A (2025) Impact of digital transformation on corporate sustainability: evidence from China's carbon emissions. Energy Inf 8(1). https://doi.org/10.1186/S42162-025-00479-8

Themelis NJ (2023) Energy and materials recovery from post-recycling wastes: WTE. Waste Dispos Sustain Energy 5(3):249–257. https://doi.org/10.1007/S42768-023-00138-2

Travassos Rosário A, Carmo Dias J (2023) The new digital economy and sustainability: challenges and opportunities. Sustainability 15(14). https://doi.org/10.3390/SU151410902

Trenerry B, Chng S, Wang Y, Suhaila ZS, Lim SS, Lu HY, Oh PH (2021) Preparing workplaces for digital transformation: an integrative review and framework of multi-level factors. Front Psychol 12.https://doi.org/10.3389/FPSYG.2021.620766/FULL

UNCTAD (2024) Digital economy report: shaping an environmentally sustainable and inclusive... digital future. United Nations. https://unctad.org/system/files/official-document/der2024_foreword_en.pdf. Accessed 25 Oct 2025

UNEP (2024) Global Resources Outlook 2024: Bend the trend—pathways to a liveable planet as resource use spikes. International Resource Panel. https://www.resourcepanel.org/reports/global-resources-outlook-2024. Accessed 25 Oct 2025

Vărzaru AA, Bocean CG (2024) Digital transformation and innovation: the influence of digital technologies on turnover from innovation activities and types of innovation. Systems 12(9). https://doi.org/10.3390/systems12090359

World Bank Group (2023) Increasing access to technology for inclusion. https://documents.worldbank.org/en/publication/documents-reports/documentdetail/099631003072338051

World Bank Group (2024) Digital progress and trends report. Washington. https://www.worldbank.org/en/publication/digital-progress-and-trends-report. Accessed 22 Oct 2025

Xia J, Wu Z, Chen B (2022) How digital transformation improves corporate environmental management: a review and research agenda. Front Environ Sci 10.https://doi.org/10.3389/FENVS.2022.943843

Zhang C, Chen P, Hao Y (2022) The impact of digital transformation on corporate sustainability—new evidence from Chinese listed companies. Front Environ Sci 10.https://doi.org/10.3389/FENVS.2022.1047418

Chapter 7
Scientific Landscapes of Circular Economy and Sustainable Consumption

Abstract This chapter explores the scientific landscape of circular economy and sustainable consumption through a bibliometric analysis of academic literature indexed in Scopus and Web of Science. It examines four strategic relationships that form the core of this book: circular economy and sustainable consumption, circular economy and SDG 12, circular economy and materials/raw materials, and circular economy and energy. Using co-occurrence mapping and cluster analysis, the study identifies dominant themes, emerging trends, and collaborative patterns that define current research priorities. Results reveal a sharp increase in publications after 2018, with peaks projected for 2024–2025, confirming the growing relevance of circular economy strategies in global sustainability agendas. Keyword networks highlight strong linkages between circular economy, sustainable development, recycling, energy efficiency, and digital transformation, reflecting a multidimensional approach that integrates technological, behavioral, and policy dimensions. Geographically, research is concentrated in Europe and Asia, led by countries such as China, Italy, Spain, and India, with notable contributions from the United States and other regions. These findings underscore the strategic role of the circular economy in shaping future sustainability beyond the 2030 Agenda. The chapter offers practical insights for educators, researchers, policymakers, and practitioners, enabling informed decisions in curriculum design, policy development, and strategic planning for circular economy transitions. By mapping conceptual structures and research dynamics, it provides a foundation for advancing knowledge and fostering innovation in sustainable production and consumption models.

Keywords Bibliometric analysis · Global research trends · Circular economy · Sustainable consumption · SDG 12 · Raw materials · Energy efficiency · Digital transformation

L. Velazquez, *Circular Economy and Sustainable Consumption in the Context of SDG 12*, Strategies for Sustainability, https://doi.org/10.1007/978-3-032-15598-6_7

7.1 Introduction

The increasing complexity of global sustainability challenges has elevated the role of the circular economy and sustainable consumption as strategic frameworks for addressing resource depletion, waste generation, and environmental degradation (Safi et al. 2023). These concepts are particularly relevant to Sustainable Development Goal number 12, which promotes responsible consumption and production patterns across sectors.

This chapter aims to provide a bibliometric perspective that complements the conceptual and theoretical discussions presented earlier in the book. It analyzes how the scientific community is shaping the future of these fields beyond current policy frameworks, offering insights into emerging trends and interdisciplinary convergence (Donthu et al. 2021). The analysis focuses on four strategic relationships that have structured the content of this book: circular economy and sustainable consumption, circular economy and SDG 12, circular economy and materials/raw materials, and circular economy and energy. These dimensions represent the core thematic pillars explored throughout previous chapters and are examined here through a global bibliometric analysis.

By mapping keyword co-occurrences, disciplinary intersections, and regional contributions, the chapter reveals a dynamic integration of environmental sciences, engineering, social sciences, and digital technologies. These findings highlight the growing relevance of topics such as Industry 4.0, energy efficiency, raw materials, and digital transformation in the context of circularity.

This chapter complements the comprehensive literature review presented throughout the book—supported by more than 800 cited references—by offering a bibliometric perspective on how circular economy and sustainable consumption are being conceptualized and operationalized across scientific domains. Through the analysis of publication trends, keyword networks, and thematic clusters, it contributes to a deeper understanding of the knowledge structures that support transformative sustainability. These insights are particularly valuable for educators designing sustainability curricula, researchers seeking emerging topics for high-impact studies, policymakers aiming to align regulations with scientific trends, and practitioners developing circular strategies in industry and public sectors.

7.2 Methodology

The bibliometric analysis was conducted using data retrieved from Scopus and Web of Science between September 15 and 17, 2025. Both databases were selected for their global coverage and complementary indexing criteria (Zhu and Liu 2020). The search strategy applied Boolean combinations of keywords across titles, abstracts, and author-defined keywords, including "Circular Economy," "Sustainable Consumption," "SDG 12," "Digital Technologies," "Raw Materials," "Energy," and "Industry

4.0." No filters were applied regarding language, document type, or disciplinary field. The retrieved records were processed using VOSviewer (version 1.6.20), applying co-occurrence mapping and association strength normalization. A minimum threshold of five keyword repetitions per category was used to ensure analytical consistency and reduce noise in the clustering process.

The analysis examined relationships among terms to identify conceptual proximity, thematic convergence, and disciplinary intersections. Co-occurrence maps revealed strong linkages between the circular economy and topics such as energy efficiency, digital transformation, and raw material management. Thematic clusters were interpreted to uncover dominant research fronts and emerging areas of inquiry. Link strength metrics and cluster density provided insights into the intensity and cohesion of these relationships. These outputs offer practical value for academic planning, particularly in guiding the selection of dissertation topics, thesis projects, and capstone initiatives. Additionally, the identification of high-impact themes and collaborative patterns supports strategic decision-making for students and educators seeking alignment with labor market trends, interdisciplinary opportunities, and future-oriented sustainability agendas.

7.3 Results

7.3.1 General Overview of Research Dynamics

The bibliometric analysis confirms the growing importance of the circular economy in global scientific discussions. Both Scopus and Web of Science show accelerated publication rates after 2018, with peaks expected around 2024–2025, indicating the establishment of the circular economy as a key theme in sustainability research. The concept of circular economy alone appears in 48,680 documents in Scopus (including 29,263 articles) and 39,820 documents in Web of Science (including 29,442 articles), emphasizing its academic significance in these major databases. Geographically, Scopus data show Italy, China, and India as leading contributors, followed by Spain, the UK, the US, and Germany, with Brazil, Portugal, and Australia also playing important roles. Conversely, Web of Science highlights China, Italy, India, and Spain at the top, along with strong participation from the US and Germany. These patterns provide a solid foundation for analyzing the four strategic relationships discussed in this chapter.

7.3.2 Global Research Trends on Circular Economy and Sustainable Consumption

The co-occurrence map (Fig. 7.1) illustrates the conceptual network linking circular economy and sustainable consumption, based on Scopus data. Both terms are positioned at the center and strongly connected to sustainable development and sustainability. The visualization reveals multiple thematic clusters that collectively represent diverse research perspectives, including behavioral aspects, policy frameworks, and operational strategies. This analysis is based on 532 documents, of which 371 are journal articles, grouped into 7 clusters—each cluster representing a set of keywords that frequently appear together, indicating thematic subfields within the research domain. The network also shows 3621 links, which represent the number of direct co-occurrence connections between keywords, and a total link strength of 9077, which measures the intensity of these connections (higher values indicate stronger conceptual relationships). In terms of frequency, circular economy appears as the most dominant term with over 350 occurrences, followed by sustainable consumption with nearly 250 occurrences, and other relevant concepts such as consumption behavior, waste management, and lifecycle. Geographically, research is led by India (69 documents), followed by the United Kingdom (65) and Italy (49), with Germany, Sweden, and China also contributing significantly, indicating strong engagement from Asia and Europe. Research activity has grown significantly in recent years, with a sharp increase after 2018 and a peak around 2024–2025, confirming that this association has become a major focus in sustainability literature.

The co-occurrence map (Fig. 7.2) illustrates the conceptual network connecting circular economy and sustainable consumption, based on Web of Science data. Both terms are prominently positioned alongside sustainable development and sustainability. This analysis is based on 3426 documents, of which 2507 are journal articles, organized into 8 clusters—each cluster representing a group of keywords that frequently appear together, signaling distinct thematic areas such as industrial strategies, consumer behavior, and resource recovery. The network includes 39,144 links, which indicate the number of direct co-occurrence connections between keywords, and a total link strength of 78,959, a measure of the intensity of these connections (higher values reflect stronger conceptual relationships and integration across topics). Within this network, circular economy is the most dominant term with over 2000 occurrences, followed by sustainability, consumption, and sustainable development, highlighting the systemic nature of research in this domain. Geographically, research is led by China, followed by Italy, India, and Spain, with strong contributions from the United States, Germany, and other European countries, indicating a global distribution with concentration in Asia and Europe. The overall growth trend shows a sharp increase after 2019, peaking in 2024–2025, reinforcing the global relevance of this research theme.

Overall, Scopus data suggest an interdisciplinary approach that combines consumer behavior and environmental economics, as reflected in its seven clusters and strong links between behavioral and policy-related concepts. In contrast, Web

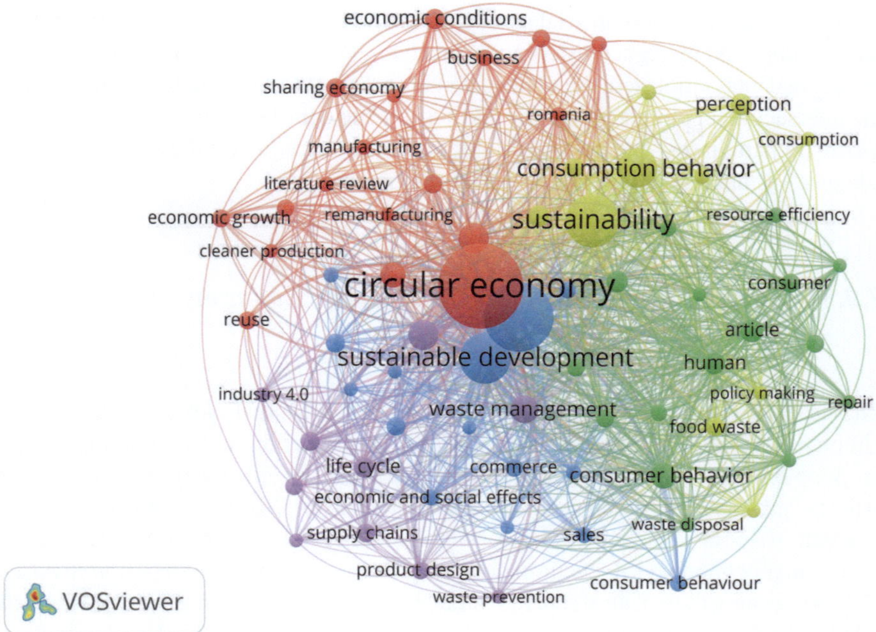

Fig. 7.1 Co-occurrence map of keywords in publications on "circular economy and sustainable consumption" based on Scopus data (September 2025). *Source* Own elaboration using VOSviewer

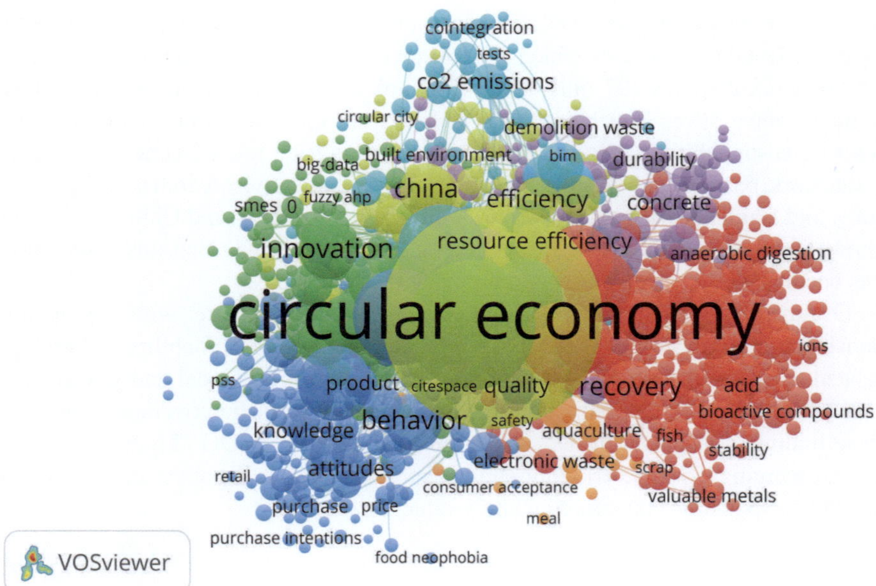

Fig. 7.2 Co-occurrence map of keywords in publications on "circular economy and sustainable consumption" based on web of science data (September 2025). *Source* Own elaboration using VOSviewer

of Science emphasizes industrial strategies, methodological frameworks, and governance, supported by eight clusters and a much higher link strength, indicating deeper conceptual integration across technical domains. Together, these findings confirm that the intersection of circular economy and sustainable consumption is evolving toward a holistic paradigm—bridging technological, behavioral, and policy dimensions to advance sustainable production and consumption models.

7.3.3 Global Research Trends on Circular Economy and SDG 12

The co-occurrence map (Fig. 7.3) illustrates the conceptual connections between circular economy and SDG 12: Responsible Consumption and Production, based on Scopus data. The analysis includes 163 documents, of which 102 are journal articles, grouped into 7 clusters. The network shows 767 links and a total link strength of 1656, reflecting strong conceptual integration. The most frequent terms are circular economy (110 occurrences), sustainable development (41), and SDG 12 (22), confirming their central role in sustainability research. Geographically, research is led by India, the United Kingdom, and Italy, with China and Spain also contributing significantly. The trend indicates rapid growth after 2021, peaking in 2024–2025, demonstrating that this association has become an emerging research priority.

The co-occurrence map (Fig. 7.4) shows the conceptual connections between the circular economy and SDG 12, based on Web of Science data. The analysis includes 169 documents, of which 133 are journal articles, grouped into 6 clusters. The network displays 497 links and a total link strength of 955, indicating strong thematic integration. The most frequent terms are circular economy (104 occurrences), sustainability (49), and SDG 12 (19), alongside related concepts such as framework, barriers, and waste management. Geographically, research is led by India, Italy, and England, with notable contributions from Germany and China. The trend shows steady growth after 2021, peaking in 2024, confirming that this association has become an emerging priority in sustainability research.

Overall, Scopus data reveal a more interconnected structure, with higher link density suggesting stronger conceptual integration among sustainability and policy-related themes. In contrast, Web of Science emphasizes technical and governance aspects, reflected in its distinct clusters and broader thematic coverage. Together, these findings confirm that research on circular economy and SDG 12 is consolidating as an emerging priority, bridging strategic, operational, and policy dimensions to advance responsible consumption and production.

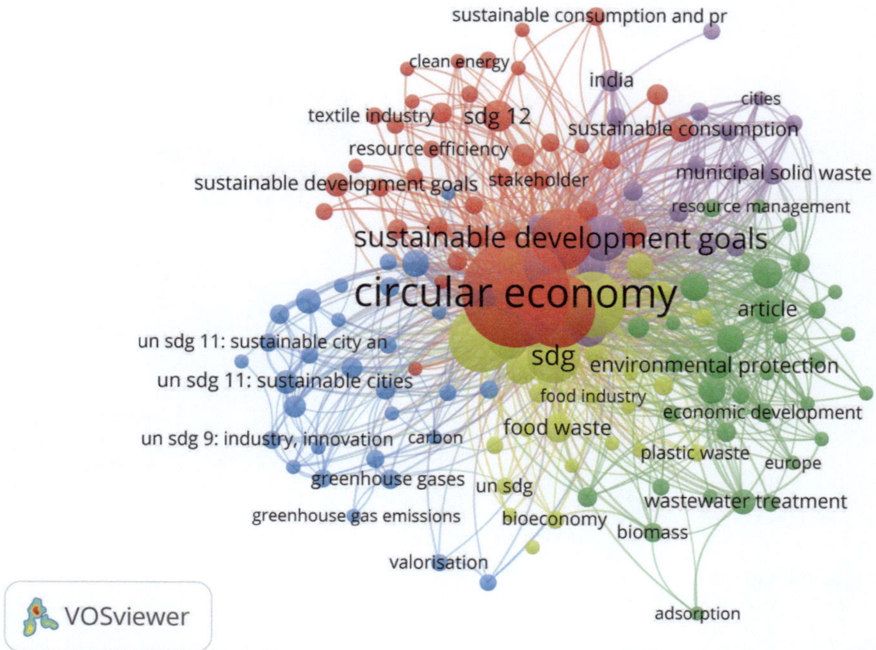

Fig. 7.3 Co-occurrence map of keywords in publications on "circular economy and SDG 12" based on Scopus data (September 2025). *Source* Own elaboration using VOSviewer

7.3.4 Global Research Trends on Circular Economy and Materials/Raw Materials

The co-occurrence map (Fig. 7.5) shows the conceptual network linking circular economy and raw materials, based on Scopus data. This analysis includes 2658 documents, with 1534 being journal articles, organized into 6 clusters, with 7690 links and a total link strength of 14,438. The most common keywords are circular economy (1697 occurrences), recycling (647), sustainable development (496), sustainability (339), and waste management (299), highlighting strong connections with resource recovery and material efficiency. Geographically, research is led by Italy, followed by Spain, Germany, and Poland, with notable contributions from the United Kingdom, Portugal, China, and India, showing a strong European presence and increasing Asian involvement. The trend over time shows little activity before 2015, followed by steady growth from 2016, speeding up sharply after 2019 and peaking between 2023 and 2024, confirming that integrating circular economy principles with raw materials has become a key research area in recent years.

The co-occurrence map (Fig. 7.6) illustrates the conceptual network linking circular economy and materials/raw materials, based on Web of Science data. This analysis includes 3103 documents, of which 2271 are journal articles, grouped into 6 clusters, with 37,288 links and a total link strength of 65,782. The most frequent

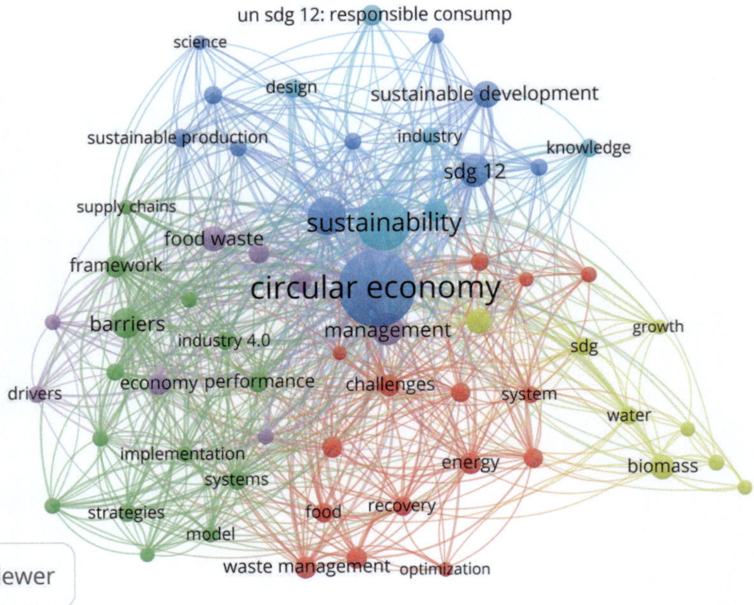

Fig. 7.4 Co-occurrence map of keywords in publications on "circular economy and SDG 12" based on web of science data (September 2025). *Source* Own elaboration using VOSviewer

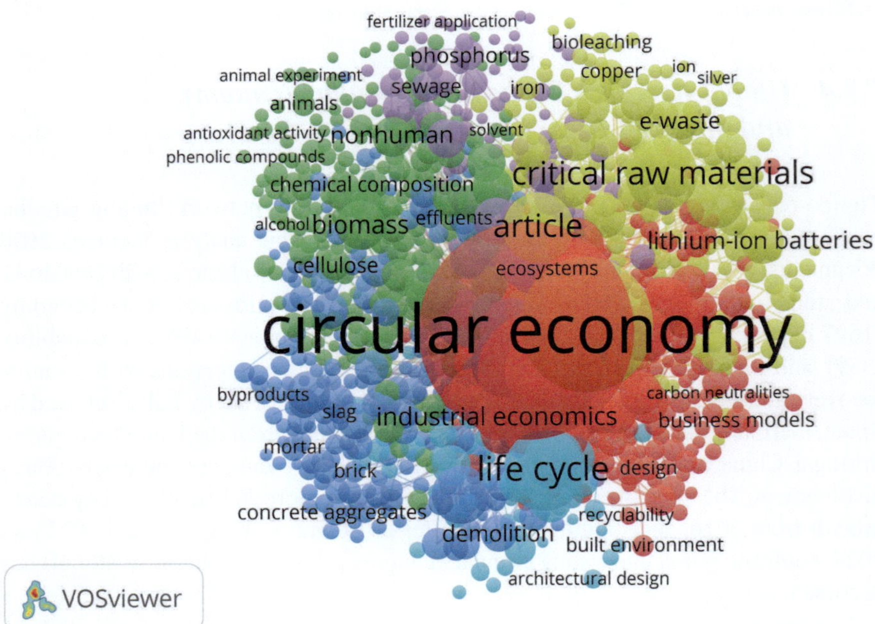

Fig. 7.5 Co-occurrence map of keywords in publications on "circular economy and materials/raw materials" based on Scopus data (September 2025). *Source* Own elaboration using VOSviewer

keywords are circular economy (1738 occurrences), sustainability (383), waste (351), recycling (331), and recovery (251), reflecting strong connections to resource efficiency and material reuse. Geographically, research is led by Italy, followed by Spain, India, and Portugal, with notable contributions from Germany, Romania, and China, indicating a concentration in Europe with growing participation from Asia. The temporal trend shows steady growth after 2016, accelerating sharply from 2019 and peaking between 2022 and 2024, confirming that the integration of the circular economy and raw materials has become a major research focus in recent years.

Scopus highlights operational themes such as recycling and waste management, showing strong links with resource recovery strategies. Web of Science, on the other hand, reflects broader conceptual integration, with higher link strength and emphasis on material efficiency and reuse. Combined, these perspectives indicate that research on circular economy and raw materials is advancing toward a comprehensive framework that connects technical innovation with sustainability objectives.

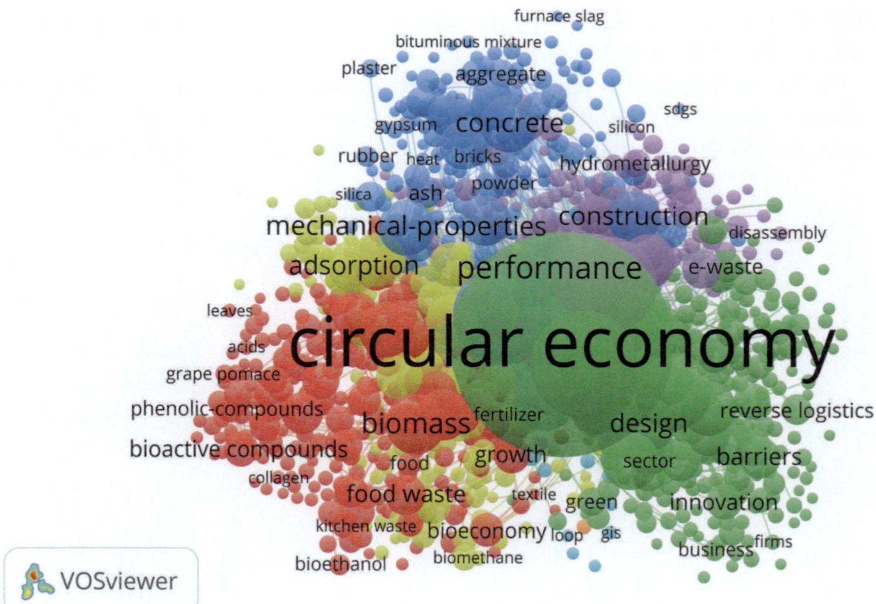

Fig. 7.6 Co-occurrence map of keywords in publications on "circular economy and materials/raw materials" based on web of science data (September 2025). *Source* Own elaboration using VOSviewer

7.3.5 *Global Research Trends on Circular Economy and Energy*

The co-occurrence map (Fig. 7.7) illustrates the conceptual connections between circular economy and energy, based on Scopus data. The analysis includes 13,217 documents, of which 7695 are journal articles, grouped into 5 clusters, with 233,496 links and a total link strength of 1,020,579, indicating extensive conceptual integration. The most frequent keywords are *circular economy* (8110 occurrences), *sustainable development* (2774), *recycling* (2098), *waste management* (1501), and *energy* (1061), reflecting strong links between energy systems, resource recovery, and sustainability. Geographically, research is led by China, followed by Italy and Spain, with significant contributions from the United Kingdom, the United States, India, Germany, and Brazil, showing a global distribution with strong European and Asian participation. The temporal trend reveals minimal activity before 2015, followed by rapid growth from 2016, accelerating sharply after 2020 and peaking in 2024–2025, confirming that energy-related circular economy strategies have become a major research focus in recent years.

The co-occurrence map (Fig. 7.8) illustrates the conceptual connections between circular economy and energy, based on Web of Science data. The analysis includes 11,666 documents, of which 8510 are journal articles, grouped into 5 clusters, with

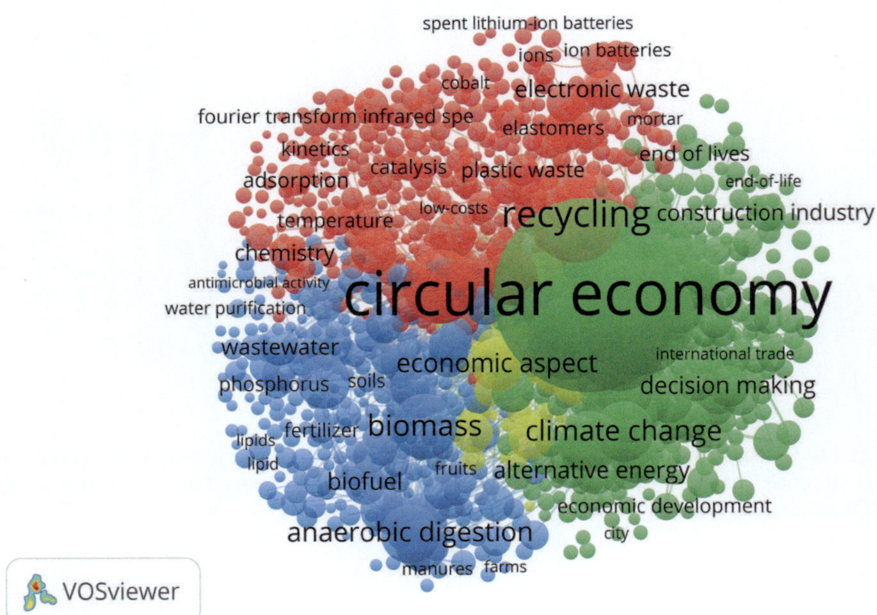

Fig. 7.7 Co-occurrence map of keywords in publications on "circular economy and energy" based on Scopus data (September 2025). *Source* Own elaboration using VOSviewer

88,540 links and a total link strength of 266,983, indicating strong thematic integration. The most frequent keywords are *circular economy* (5893 occurrences), *energy* (1825), *sustainability* (1614), *lifecycle assessment* (1008), and *management* (932), reflecting a focus on energy systems, performance, and governance within circular strategies. Geographically, research is led by China, followed by Italy, Spain, and India, with notable contributions from the United States, Germany, and South Korea, showing a global distribution with strong European and Asian participation. The temporal trend shows minimal activity before 2015, followed by rapid growth from 2016, accelerating after 2020 and peaking in 2024, confirming that energy-related circular economy research has become a major focus in recent years.

Scopus data reveal a broader scale and higher link strength, emphasizing operational themes such as recycling, waste management, and energy recovery within circular strategies. In contrast, Web of Science highlights governance and performance aspects, integrating concepts like lifecycle assessment and management alongside energy systems. Together, these perspectives confirm that energy-related circular economy research is evolving toward a multidimensional framework, bridging technical innovation, sustainability, and policy considerations.

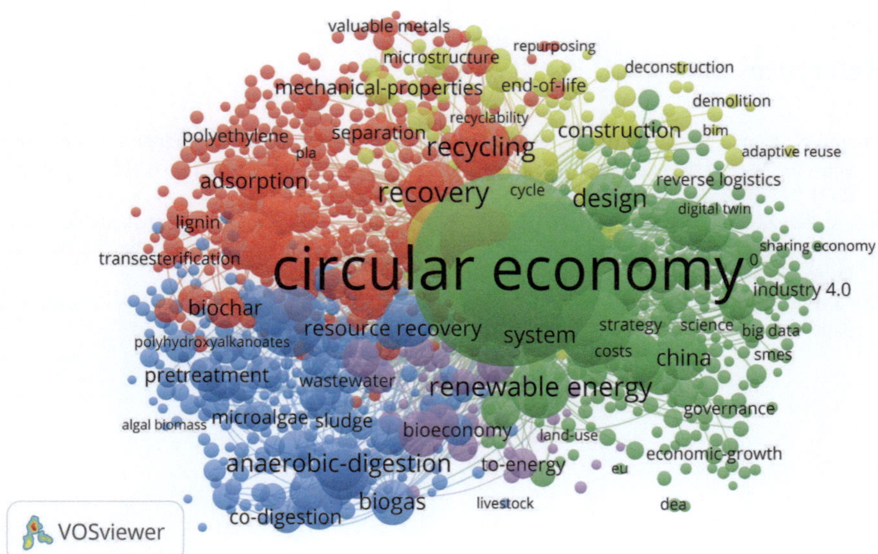

Fig. 7.8 Co-occurrence map of keywords in publications on "circular economy and energy" based on web of science data (September 2025). *Source* Own elaboration using VOSviewer

7.4 Conclusion

The comparative analysis of Scopus and Web of Science data reveals both convergence and divergence in research priorities related to the association between the circular economy and sustainable consumption. Across all thematic areas—SDG 12, raw materials, and energy—the findings confirm accelerated growth after 2018, with peaks projected for 2024–2025. Europe and Asia dominate the research landscape, supported by strong interdisciplinary integration of environmental sciences, engineering, and digital technologies.

These patterns indicate that future research will increasingly align with digitalization, energy transition, and systemic policy frameworks, shaping sustainability beyond the 2030 Agenda. By examining the four strategic relationships that structure this book, the chapter provides actionable guidance for educators, researchers, policymakers, and practitioners. These insights support curriculum development, inform evidence-based policy design, and enable organizations to anticipate technological and regulatory shifts that will shape circular economy implementation in the coming decades.

References

Donthu N, Kumar S, Mukherjee D, Pandey N, Lim WM (2021) How to conduct a bibliometric analysis: an overview and guidelines. J Bus Res 133:285–296. https://doi.org/10.1016/j.jbusres.2021.04.070

Safi A, Wei X, Sansaloni EM, Umar M (2023) Breaking down the complexity of sustainable development: a focus on resources, economic complexity, and innovation. Resour Policy 83:103746. https://doi.org/10.1016/J.RESOURPOL.2023.103746

Zhu J, Liu W (2020) A tale of two databases: the use of web of science and scopus in academic papers. Scientometrics 123(1). https://doi.org/10.1007/s11192-020-03387-8

Epilogue

Beyond 2030: Reflections on the Future of Circular Economy and Sustainable Consumption

As the global community approaches the horizon of the 2030 Agenda for Sustainable Development, students and emerging professionals in sustainability face a pivotal moment. Whether months away from entering the workforce, already navigating early career paths, or bringing prior experience into new contexts, each must confront a fundamental question: What kind of professional do I want to be—and what kind of professional do productive sectors now require? The principles of circular economy and sustainable consumption, thoroughly explored throughout this book, are no longer optional competencies; they are essential foundations for shaping resilient, equitable, and regenerative systems. In this evolving landscape, technical expertise must be matched by ethical clarity, systemic thinking, and a commitment to long-term transformation.

The 2030 Agenda for Sustainable Development, initially embraced as a global framework for coordinated action, now approaches its conclusion amid turbulence and uncertainty. Geopolitical tensions, armed conflicts, and shifting national priorities have eroded the momentum that once mobilized governments, industries, and academia. Although Circular Economy and Sustainable Consumption were conceptualized long before the Agenda's launch in 2015, their formal inclusion in SDG 12 marked a turning point in circularity discourse. Yet, with several targets unlikely to be fully met, and critiques framing the Agenda as aspirational rather than actionable, the challenge is not to lament its shortcomings—but to build upon its foundations. In this spirit, the recent adoption of the Pact for the Future could be a renewed multilateral effort that future professionals may leverage to accelerate sustainability commitments and address emerging global challenges beyond 2030.

As concerns about resource scarcity, environmental pollution, and climate change have intensified, industries have begun adopting circular economy and sustainable consumption strategies—such as product take-back programs, material recycling,

L. Velazquez, *Circular Economy and Sustainable Consumption in the Context of SDG 12*, Strategies for Sustainability, https://doi.org/10.1007/978-3-032-15598-6

remanufacturing, energy recovery from industrial byproducts, and the integration of smart energy grids to optimize resource use. This book has presented diverse examples that illustrate how industrial sectors and other domains—including construction, agriculture, and services—are integrating innovative solutions to achieve circularity in both material and energy systems. These efforts reflect a broader transformation that began well before the 2030 Agenda and will likely continue beyond it, driven by the need to reconcile economic activity with ecological limits. In this context, circularity is not a temporary trend but a long-term strategic imperative.

While previous generations have laid the scientific and conceptual groundwork for circularity—from the 1960s to today—current and future professionals are called to lead its full-scale implementation. The responsibility now shifts to those who will transform industries, design regenerative systems, and develop innovative solutions that balance production and consumption with planetary boundaries. The principles of circular economy, once outlined by pioneers and further refined by those who followed, must now be turned into practical applications across various sectors. This task demands more than technical skill; it requires a new generation equipped with systems thinking, ethical leadership, digital fluency, and the ability to drive change in complex, uncertain environments.

MIX
Papier aus verantwortungsvollen Quellen
Paper from responsible sources
FSC® C105338
www.fsc.org

If you have any concerns about our products,
you can contact us on
ProductSafety@springernature.com

In case Publisher is established outside the EU,
the EU authorized representative is:
Springer Nature Customer Service Center GmbH
Europaplatz 3, 69115 Heidelberg, Germany

Printed by Libri Plureos GmbH
in Hamburg, Germany